PHYSICS

IN 50 MILESTONE MOMENTS

METRO BOOKS
New York

An Imprint of Sterling Publishing Co., Inc.
1166 Avenue of the Americas
New York, NY 10036

ISBN: 978-1-4351-6475-8

For information about custom editions, special sales, and
premium and corporate purchases, please contact
Sterling Special Sales at 800-805-5489 or
specialsales@sterlingpublishing.com.

Manufactured in China

1 3 5 7 9 10 8 6 4 2

www.sterlingpublishing.com

Design and illustration by Simon Daley at Giraffe Books

PHYSICS

IN 50 MILESTONE MOMENTS

A TIMELINE OF SCIENTIFIC LANDMARKS

JAMES LEES

METRO BOOKS

CONTENTS

INTRODUCTION

Physics is incredibly important to our modern lives. It has led us to numerous remarkable achievements, including such trivial matters as flight, computers, and the smashing together of the tiniest things in the universe to draw power from them. Needless to say, arriving at this point has not been easy; physics is an endeavor that has been some 10,000 years in the making.

In this book we take you through 50 of the most significant milestones in physics, all of which helped us get to where we are today, taking in the work of the ancient Greeks and of greats such as Newton and Einstein, and moving on to some of the most exciting experiments being conducted today. Moving through the entries, it will quickly become evident that the milestones do not exist in isolation; each one is a single point within the ever-changing story of scientific progress. After all, progress in physics (and by extension in all of science) is marked by the often incremental acquisition of knowledge, by scientists who build on the work of their predecessors.

This book is not a definitive account by any means—it could contain 500 milestones and still not lay claim to being definitive—and the selection of the milestones is a subjective process. Nevertheless, the entries presented here have all made a huge contribution to the evolution of physics, and their effects are still felt today.

A NOTE ON DATES

Many of the dates from the period up to the Enlightenment era are a best guess. This is because sources from that period can be fairly unreliable and they were often not dated. Additionally, prior to 1752, it wasn't unusual for a source to be given two dates—true of some of the sources for this book. The reason for this lies in the old Julian calendar, which held a year to be exactly 365.25 days, which was less precise than its successor, the Gregorian calendar, which has a year as 365.2425 days. By the time of Pope Gregory XIII (1572–1585), the accumulative effect of this imprecision could be seen in the changing of the seasons, which were off by 10 days. To rectify this, Pope Gregory produced his eponymous calendar. However, adoption of this was slow; it wasn't formally adopted in Britain until 1752, which meant that some sources prior to this carried Julian and Gregorian dates. This book uses the Gregorian calendar where possible.

SPACE AGE: New technology, including the Hubble Space Telescope (see pages 168-171), has revealed to us for the first time the enormity of the universe.

MILESTONE PHYSICISTS

NAME	DATES	NATIONALITY	OUTSTANDING ACHIEVEMENT	PAGE
Al-Hasan Ibn Al-Haytham	965-1040	Persian	Describing light	32
Archimedes	ca. 287-212 BC	Sicilian	Displacement theory	22
Aristotle	ca. 384-ca. 322 BC	Greek	Aristotelian logic	18
Bardeen, John	1908-1991	American	Development of the transistor	146
Bell, Alexander Graham	1847-1922	Scottish	Invention of the telephone	100
Bohr, Niels	1885-1962	Danish	Explaining spectral lines	120
Boltzmann, Ludwig	1844-1906	German	Creation of statistical mechanics	92
Brahe, Tycho	1546-1601	Danish	Accurate and comprehensive astronomical observations	40
Brattain, Walter	1902-1987	American	Development of the transistor	146
Carnot, Sadi	1796-1832	French	Developing thermodynamics	82
Copernicus, Nicolaus	1473-1543	German	Challenging the Ptolemaic system	36
Dalton, John	1766-1844	English	Developing atomic theory	80
Einstein, Albert	1879-1955	German	Relativity, quantum mechanics, Brownian motion, and mass-energy equivalence	112 & 124
Euler, Leonhard	1707-1783	Swiss	Euler's Identity	64
Fahrenheit, Daniel	1686-1736	German	Creating a standardized temperature scale	58
Faraday, Michael	1791-1867	English	Discovery of electromagnetic induction	86
Feynman, Richard	1918-1988	American	Development of Feynman diagrams	150
Galilei, Galileo	1564-1642	Italian	Discovering the moons of Jupiter	48
Geiger, Hans	1882-1945	German	Completing the description of the atom	116
Gell-Mann, Murray	1929-	American	Proposal of subatomic particles	160

NAME	DATES	NATIONALITY	OUTSTANDING ACHIEVEMENT	PAGE
Gödel, Kurt	1906-1978	Austro-Hungarian	Creation of the incompleteness theorem	134
Goodricke, John	1764-1786	English	Explaining Cepheid variable stars	68
Halley, Edmond	1656-1742	English	Predicting the return of Halley's Comet	66
Hamilton, William	1805-1865	Irish	Developing Hamiltonian mechanics	90
Heisenberg, Werner	1901-1976	German	Creation of the uncertainty principle	128
Hubble, Edwin	1889-1953	American	Discovering universal expansion	132
Ladislaus the Posthumous	1440-1457	Hungarian	Introduction of Arabic numbers to the West	34
Lippershey, Hans	1570-1619	Dutch	Inventing the telescope	44
Marsden, Ernest	1889-1970	English	Completing the description of the atom	116
Maxwell, James Clerk	1831-1879	Scottish	The Maxwell equations	96
Michelson, Albert	1852-1931	Prussian	Disproving the theory of Luminiferous ether	102
Morley, Edward	1838-1923	American	Disproving the theory of luminiferous ether	102
Newton, Sir Isaac	1642-1726	English	Theory of universal gravitation	54
Planck, Max	1858-1947	German	Theory of the quantization of energy	108
Ptolemy	ca. 100-ca. 170	Egyptian (probably)	Ptolemaic system	26
Rutherford, Ernest	1871-1937	New Zealander	Discovery of the types of radiation	116
Schrödinger, Erwin	1887-1961	Austrian	Creation of the Schrödinger equation	140
Shockley, William	1910-1989	American	Development of the transistor	146
Young, Thomas	1773-1829	English	Discovering the wave nature of light	76
Zwicky, Fritz	1898-1974	Bulgarian	Proposing dark matter	136

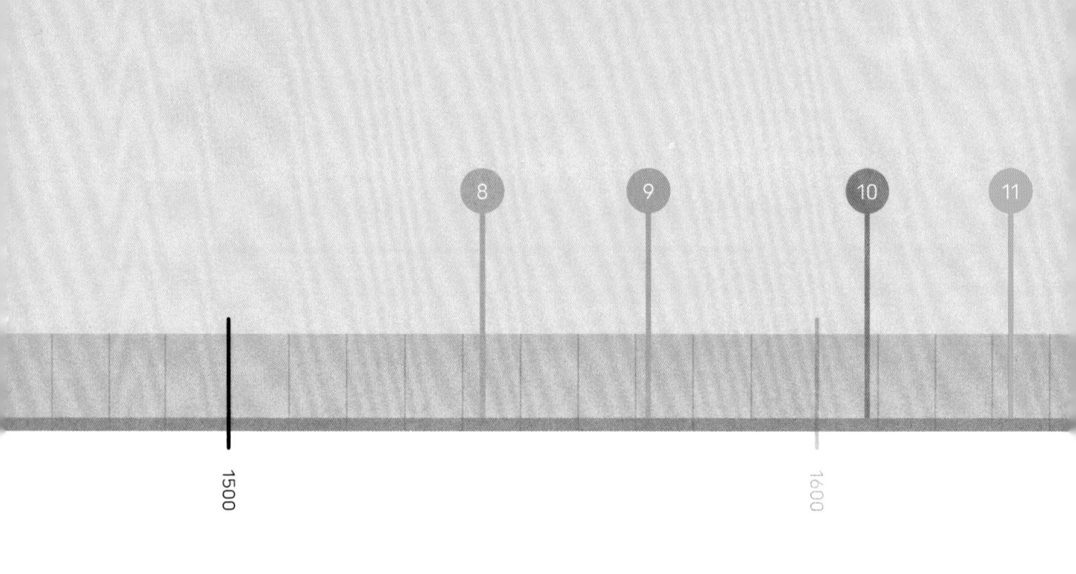

13 1687 .... PAGE 55
ISAAC NEWTON PUBLISHES *PRINCIPIA*
Newton cements in place the modern scientific method, championing empiricism, and offering an explanation for much of the world around us.

14 1714 .... PAGE 58
DANIEL GABRIEL FAHRENHEIT INVENTS THE MERCURY THERMOMETER Fahrenheit produces a thermometer that is still used in many homes today, and with it the first standard temperature scale.

15 1748 .... PAGE 64
EULER'S IDENTITY IS PUBLISHED Euler's Identity shows us the inherent interconnectedness of mathematics.

16 1759 .... PAGE 66
HALLEY'S COMET RETURNS AS PREDICTED The work of English astronomer Edmond Halley (1656–1742) enables the comet's return to be accurately predicted for the first time.

17 1784 .... PAGE 68
JOHN GOODRICKE EXPANDS THE HEAVENS
Goodricke's discovery of orbiting stars expands our understanding of the size of our universe.

18 1795 .... PAGE 72
THE METRIC SYSTEM IS INTRODUCED IN FRANCE
A standardized system of measurements facilitates the collaboration of scientists around the world.

19 1797 .... PAGE 74
HENRY CAVENDISH CALCULATES G A century after Newton defined gravity as the unknown constant G, Henry Cavendish calculates it.

20 1803 .... PAGE 77
YOUNG'S DOUBLE-SLIT EXPERIMENT IS PERFORMED Thomas Young (1773–1829) creates an experiment that will later prove light is a wave.

21 1804 .... PAGE 80
JOHN DALTON DEVELOPS ATOMIC THEORY
Dalton takes the first big step in understanding the properties of atoms.

22 1824 .... PAGE 83
CARNOT DESCRIBES THE PERFECT ENGINE
Sadi Carnot, the "father of thermodynamics," provides the key to harnessing the potential of steam engines.

23 1831 .... PAGE 86
MICHAEL FARADAY CREATES THE FARADAY DISK Faraday invents the first electrical generator and provides the means by which the modern world is powered.

24 1833 .... PAGE 90
HAMILTONIAN MECHANICS IS CREATED
William Hamilton introduces a system of mathematics that transforms the way physicists approach their subject.

PHYSICS TIMELINE

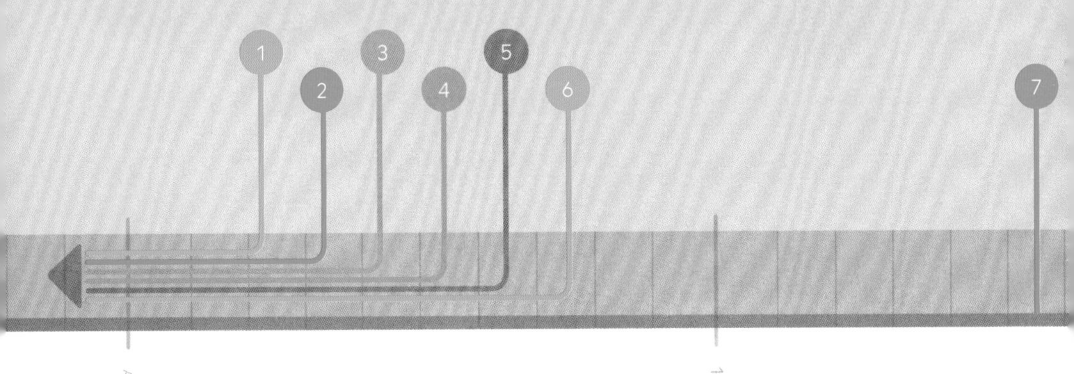

1 CA. 8000 BC .... PAGE 14
THE BUILDING OF THE WARREN FIELD CALENDAR The Warren Field calendar's discovery shows that humans began measuring the sky some 5,000 years earlier than we originally thought.

2 CA. 2300 BC .... PAGE 16
THE WRITING OF THE *ENUMA ANU ENLIL* The Babylonians attempt to chart the night sky in a collection of tablets known as the *Enuma Anu Enlil*.

3 CA. 350 BC .... PAGE 18
ARISTOTELIAN LOGIC TAKES ROOT Classical modes of thinking about the world are transformed by Aristotle's ideas.

4 CA. 25 BC .... PAGE 23
ARCHIMEDES PROBABLY DOESN'T TAKE A BATH Early physics is moved forward through the simple application of mathematics.

5 CA. AD 150 .... PAGE 26
PTOLEMY PLACES US AT THE CENTER OF THE UNIVERSE Ptolemy seeks to find mathematical proof of the geocentric model, and produced an astronomical system that lasted for 1,000 years.

6 CA. 1015 .... PAGE 32
IBN AL-HAYTHAM DESCRIBES LIGHT The work of Hasan Ibn al-Haytham (965-1040) helps begin the scientific revolution that would change the world.

7 1456 .... PAGE 34
LADISLAUS THE POSTHUMOUS SENDS A LETTER USING ARABIC NUMBERS Arabic numerals form the basis of the number system used today.

8 1543 .... PAGE 36
NICOLAUS COPERNICUS PUBLISHES *DE REVOLUTIONIBUS* Copernicus revolutionizes the way humans view their place in the universe.

9 1572 .... PAGE 40
A NEW STAR APPEARS IN THE SKY In 1572 a bright new star appears in the constellation of Cassiopeia, and once again the foundations of astronomy are shaken.

10 1608 .... PAGE 44
HANS LIPPERSHEY DESIGNS THE FIRST PRACTICAL TELESCOPE The invention of the first practical telescope dramatically broadens our scientific horizons.

11 1633 .... PAGE 48
GALILEO IS TRIED FOR HERESY Galileo Galilei's revolutionary ideas place him at odds with the Catholic Church.

12 1660 .... PAGE 52
THE ROYAL SOCIETY IS FOUNDED The Royal Society of London helps set the standard for scientific research through grants, training, and international cooperation.

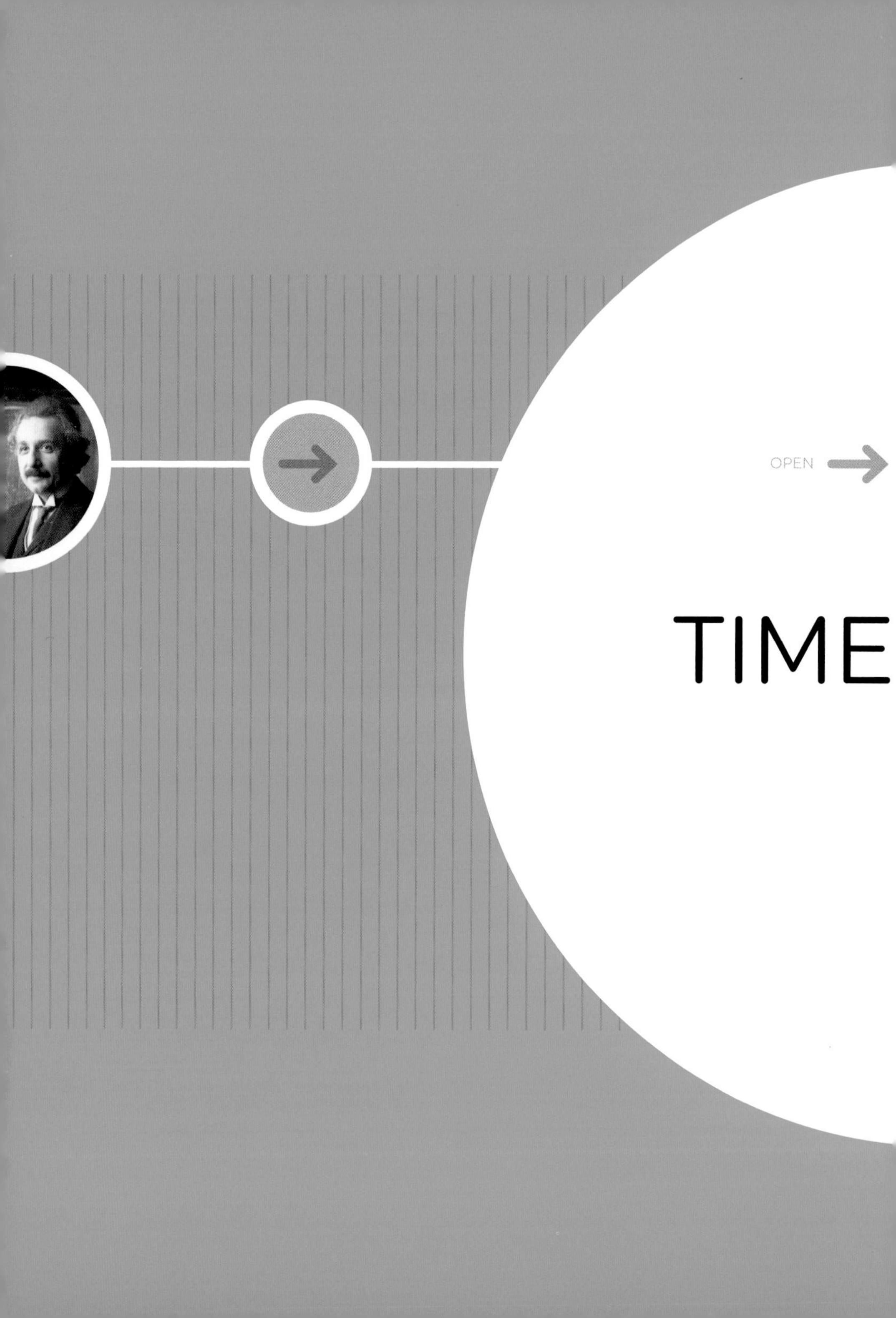
OPEN
TIME

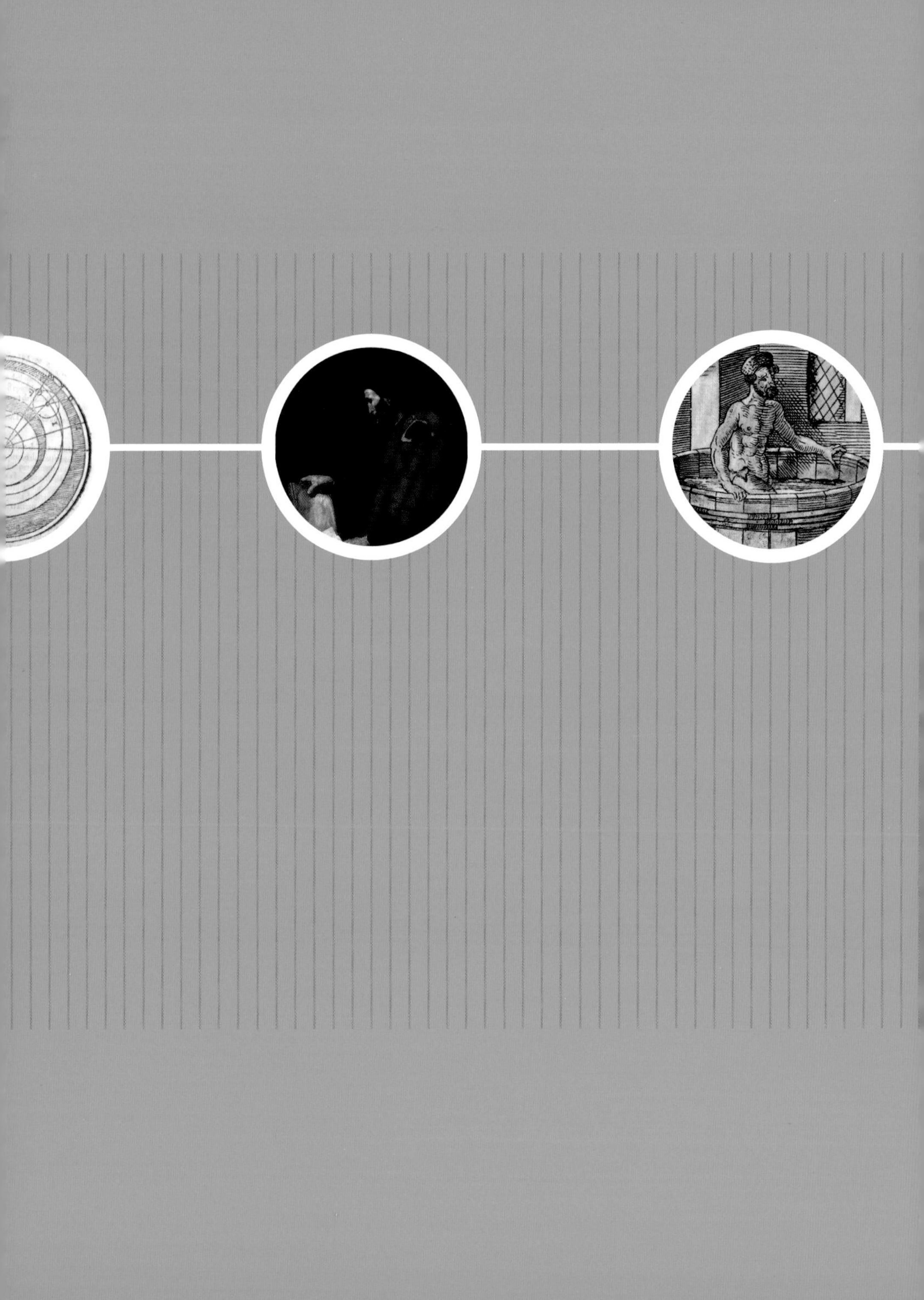

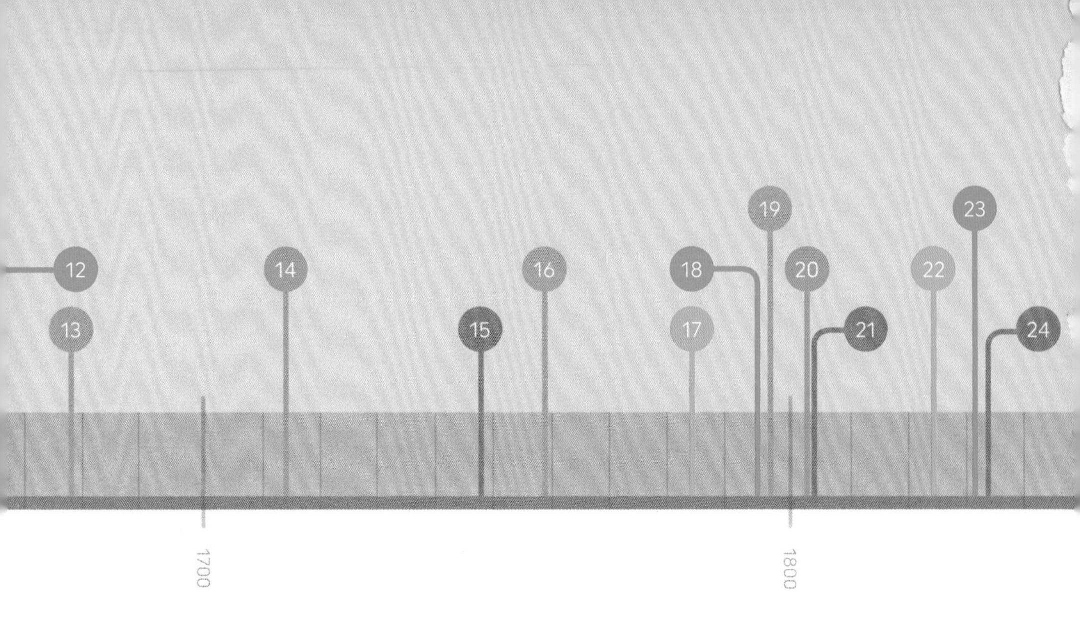
12
13
14
15
16
17
18
19
20
21
22
23
24
1700
1800

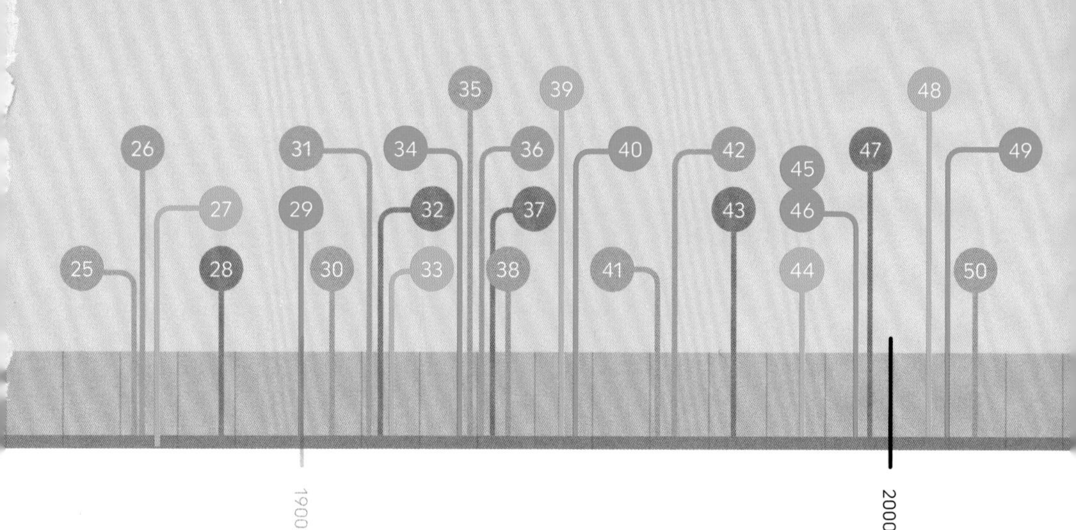
25
26
27
28
29
30
31
32
33
34
35
36
37
38
39
40
41
42
43
44
45
46
47
48
49
50
1900
2000

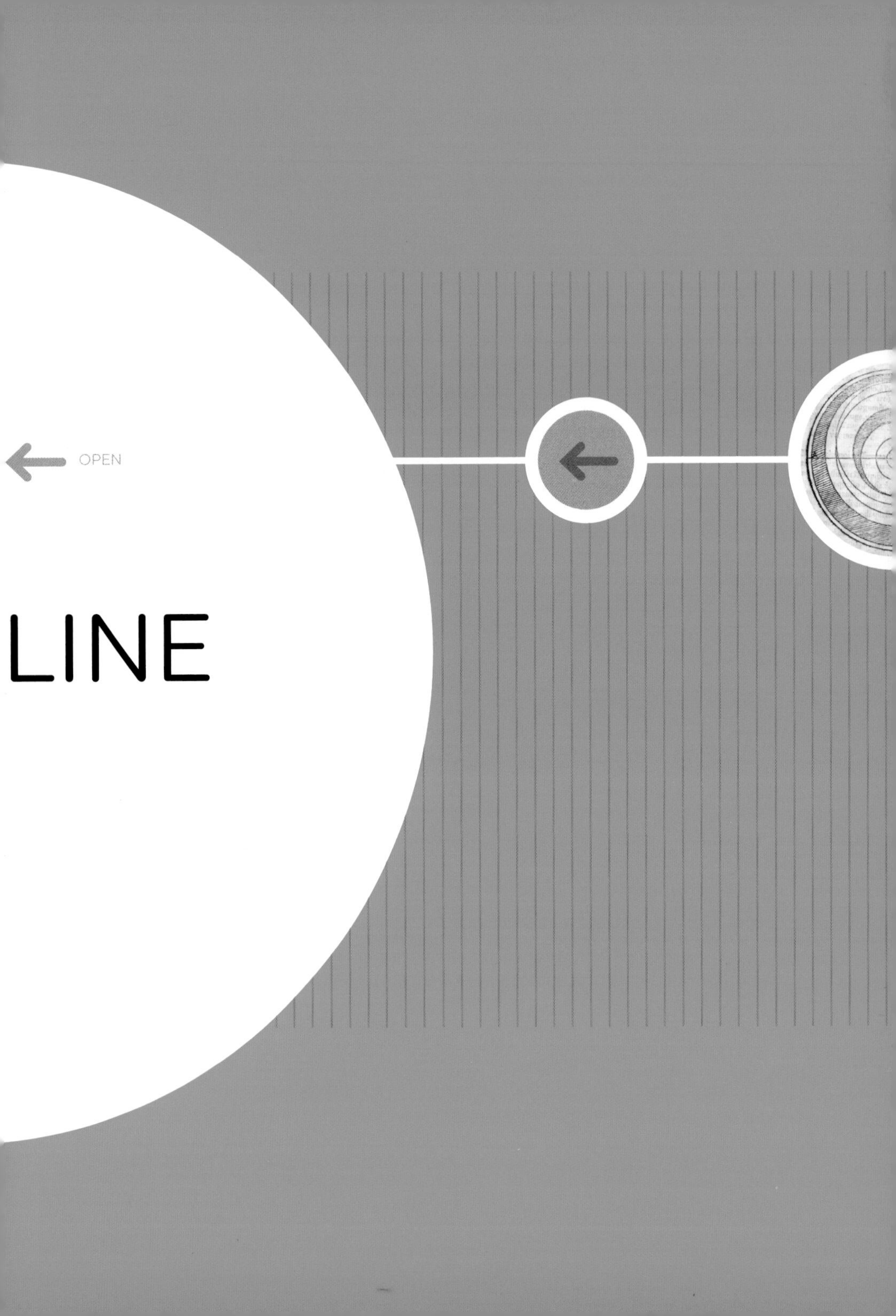
OPEN
LINE

1

THE ANCIENTS

THE BUILDING OF THE WARREN FIELD CALENDAR

Much of physics in ancient times was focused on astronomy, and if you look out into a clear night sky it's not hard to understand why. In 2013 the discovery of the Warren Field calendar showed that humans began measuring the sky some 5,000 years earlier than was originally thought.

Understanding time is very important. In today's world, there are deadlines, timetables, and plenty of other things that keep us constantly checking our watches. Ten thousand years ago, however, time could be a matter of life and death; knowing when seasonal migrations would happen or when certain plants would bear fruit could mean survival for a hunter–gatherer. And before the advent of clocks, the night sky represented a means of charting the passage of time.

The Warren Field calendar consists of twelve pits, aligned along a rough arc, that were dug near Crathes Castle in Scotland. It is thought that by standing in a specific place and observing where the moon rises in relation to the pits, it is possible to track the moon over a lunar month, as it progresses through the lunar phases (see right) in a period of about 29.5 days. That there are twelve pits also suggests that the calendar can be used to track the lunar months. By checking the position and phase of the moon, it can also provide a rough estimate of the day of the year, as well as allow longer periods of time to be gauged (ie, the time remaining in the year).

The difficulty with a lunar calendar is that a lunar year lasts 354 days, whereas the natural

EARLIER INSTRUMENTS?

The Warren Field calendar is the oldest known scientific creation in the world by a long way. It was ancient even when the Great Pyramids of Giza and Stonehenge were being constructed, and it is nearly four times farther back in history than the peak of the Roman Empire. But it is possible that the Warren Field calendar is not the first of its kind. Some have suggested that marks on a bone baton from 25,000 BC or some of the images in the Lascaux Caves from around 15,000 BC could also be lunar calendars. But such claims have proven controversial, and the Warren Field calendar is generally accepted as the first.

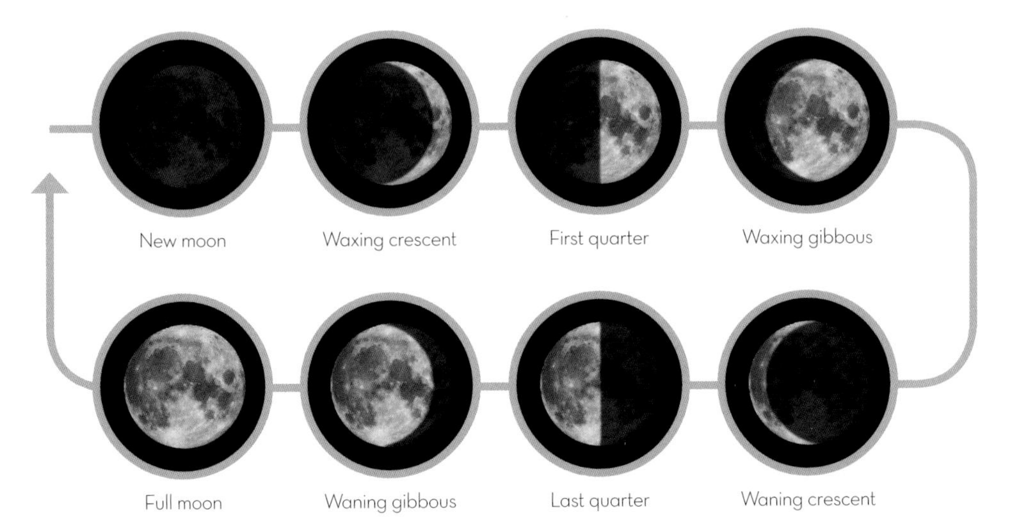

PHASES OF THE MOON: The phases of the moon, which Mesolithic-era humans were able to chart with the help of the Warren Field calendar.

(or solar) year is 365.25 days long. This means that any key times of the year you wished to mark, and therefore anticipate in the future, would very quickly start to move down your calendar—something that is called seasonal drift. Unless the calendar were adjusted each year, it would quickly become useless. Amazingly, not only is the Warren Field calendar the earliest known scientific instrument, it also gives us the first example of calibration. Each year, during the winter solstice, the calendar would be "reset" by digging the pits again, based on the position of the sunrise, thus ensuring it remained accurate for the year ahead.

WHY CREATE A CALENDAR?

The exact use of the Warren Field calendar is not known. It may have been built to help prepare for the coming of fish or migratory animals, which would be hunted for food. Alternatively, it might have had a more spiritual purpose, allowing for significant dates to be tracked accurately. Or perhaps it was a tool for interpreting the meaning of the moving heavens. Whatever its actual purpose, the calendar shows how early concepts of time and science were beginning to evolve.

What is even more important is that they *built* something. In addition to their ancestors' mode of sharing of knowledge orally, these early scientists created a scientific instrument—a key stage in the development of physics. Every experiment that has ever occurred, and every discovery and invention, has required some form of instrument. The fact that the Warren Field calendar is widely recognized as the first known scientific instrument (see box, left) secures its place in the history of science.

THE WRITING OF THE *ENUMA ANU ENLIL*

While the Babylonians may not have been the first to try to make sense of the night sky, they did make it the center of their religious customs, and they made an effort to document the heavens. This is first recorded in a collection of tablets known as the *Enuma Anu Enlil*, the first of which dates back to ca. 2300 BC.

This collection of 70 stone tablets is named for the Babylonian gods Anu and Enlil (the gods of sky and wind, respectively). Together, they contain around 7,000 entries. The tablets are in one respect a Babylonian equivalent of the Bible, in the sense that the texts they contain interpret the will of the gods and record celestial events.

Such recording of information was important because it provided a historical record of previous celestial events, which could be referred to when interpreting events in the present. It also meant that smaller, less notable events from earlier generations, which may not have been preserved by oral tradition, were available to scholars. While this method of inscription was used largely in astrology for divining purposes, it became a keystone of astronomy. The *Enuma Anu Enlil* text itself exists now mostly as fragments and to this day has not been fully translated from Cuneiform. Some of the tablets describe patterns in the moon's motion and even begin to predict lunar eclipses. The tablets concerning the sun have largely been destroyed or are missing. The translations of the texts about the planets and stars, though somewhat unreliable, appear to predict planetary and stellar motion, and even include instructions for finding celestial objects using an astrolabe (an instrument that can calculate the height of objects from the horizon) and a star catalog.

THE FIRST STAR CATALOGS

The Babylonians created the first catalogs listing constellations, individual stars, and planets, the two most important of which are *Three Stars Each* and *MUL.APIN*. Early star catalogs created many of the modern stellar constellations and set out some of the important structures, such as the zodiac (although with eighteen constellations, as opposed to the modern twelve). These Babylonian catalogs were later adopted by the Greeks and Egyptians, and provided the foundations of much of their astronomy.

Crucially, there wasn't just one set of *Enuma Anu Enlil* tablets. Multiple sets were produced, including a luxury 16-tablet edition on ivory board, and sent across the kingdom, thus allowing knowledge to be shared. The distribution of the tablets helped ensure that scientific knowledge was for one of the first times no longer exclusive to the king's court or high temple.

REGULAR RECORDS

Daily recording of astronomical objects is now commonplace, from the accurate measurement of star positions to observations of the sun, as carried out by NASA's SOHO spacecraft. But the Babylonians were the ones who began a daily record of the events, such as the following:

*Year 97, month IX, night of the 1[3th?....] ....,
measured;
the bright star of the Old Man stood in
culmination,
lunar eclipse; on the east side
when it began, in 21 of night all of it became
covered;
16 of night totality; when it began to clear,
it cleared in 19 of night from east and north to
the west?; 56 onset, totality,
[and clear]ing; at one-half beru after sunset.
[....] eclipse; in its eclipse, Sirius*

In general terms, the above translation explains that on the thirteenth night of month 9, year 97 (by the Babylonian calendar) the constellation "Old Man" was at its highest point in the night sky and that the moon eclipsed.

CHARTING VENUS The Venus Tablet of Ammisaduqa, the 63rd tablet of the *Enuma Anu Enlil*. It records the rising and setting of the planet Venus over 21 years.

Records like this were made each day. Much like astronomy today, many of the dates simply read that it was cloudy or rainy and that observation wasn't possible. For example:

Ni[ght] of the 14th, sunset to moonrise: 8° 20′; clouds, I did not watch; very overcast. The 14th, all day clouds crossed the sky. Night of the 15th, clouds crossed the sky, slow rain.

ARISTOTELIAN LOGIC TAKES ROOT

Reasoning and logic in science are things we now take for granted. In Greek times, however, things were very different. Ideas such as form, goodness, and evil lay at the heart of everything, along with the everpresent explanation that "the Gods did it." But Aristotle (ca. 384–ca. 322 BC) and his teachings turned all of this on its head.

Aristotle was a student of Plato, and after Plato's death in 347 BC he moved to Macedonia to teach Alexander the Great. Here he began a study of empiricism—the theory that all knowledge is derived from experience based on the human senses. After returning to Athens, he opened a school, the Lyceum.

In his works, Aristotle set out his views on logic (also referred to as "term logic" and "syllogistic logic"), a form of deductive reasoning that would remain dominant until the late 1800s (see the next section). Among the most prominent of the works is *Organon*, which comprises six books that fully explain his ideas of logic. These became incredibly popular, thanks partly to the profile he gained as a student of Plato and teacher to Alexander the Great.

Aristotle also set out the laws of noncontradiction and the excluded middle. The first of these simply states that no proposition can be both true and false, and the second follows this up by stating that a proposition must be either true or false. While this may seem obvious to us, to Greeks such things were not so certain. In a world of mysticism and the divine will, for things to be both true and false, or somewhere in the middle, was not entirely out of the question.

HOW ARISTOTELIAN LOGIC WORKS

One major type of Aristotelian logic is term logic, a form of deductive reasoning. Put simply, deductive reasoning involves taking a number of premises and then drawing a conclusion. It states that, if the premises are true, the conclusion must also be true. To take a famous example, look at the following two premises:

All men are mortal.
Socrates is a man.

From these two initial premises, we draw the conclusion:

Socrates is mortal.

From a position of absolute ignorance, it isn't possible to state whether Socrates is mortal or immortal. However, by looking at the two premises—that he is a man and that all men are mortal—and by accepting that they are

FIRST TEACHER: A painting of Aristotle from 1637. Thanks to his enormous contributions to many different fields and his tutelage of Alexander the Great, he is often called "The First Teacher."

both true, we can deduce that he is indeed mortal. The beauty of this kind of logic is that, if the premises are true (verifying this isn't as easy as you might expect), then so must the conclusion also be true. And we can reach these conclusions without additional external knowledge or reasoning.

This idea can be extended beyond the specific terms of "Socrates" and "mortality":

All As are Bs
All Cs are As

Therefore:

All Cs are Bs

You can use anything in place of A, B, and C. As long as the first two statements are true, the third must also be true. The benefits that this form of logic conveys are significant; it allows us to reach conclusions about things we have no knowledge about from things that we do. While this perhaps seems fairly obvious today, it was a huge breakthrough at the time.

Aristotle also worked on inductive reasoning, which involves using particular premises to reach a universal conclusion. For example:

This copper bar is metal and conducts electricity
This iron bar is metal and conducts electricity
This steel bar is metal and conducts electricity

Therefore:

All metal bars conduct electricity

Inductive reasoning is a useful method, but it is not a fail-safe way of determining the truth. For example, it could allow me to conclude that all metals are solid, because all my metal bars are solid. But the existence of a metal such as mercury, which is liquid at room temperature, shows this to be false, as does the fact that a metal will become liquid at a certain temperature. Because of this, Aristotle called inductive reasoning a secondary form of logic, though its use is still widespread, and it is often employed to make reasonable assumptions about what will happen in experiments before they are performed.

THE RISE OF EMPIRICISM

Empiricism is the theory that all knowledge is based on experience derived from the senses. In Aristotle's time, this was at odds with Platonism, which separated knowledge into various realms, including the physical and the mental, and also incorporating the divine realm. The emergence of empiricism shifted the focus onto tangible things that could be measured and tested, and it has remained fundamental to our current ideas about knowledge.

It wasn't until the 17th and 18th centuries that empiricism became established as a major branch of philosophy. Nonetheless, almost all science, even before this, was rooted in the empirical school of thought, which was based on the works of Aristotle.

THE REACH OF HIS WORK

Aristotle managed to touch upon just about every field of knowledge during his lifetime, from astronomy to zoology, and he made a significant contribution to their development. His Lyceum school helped to establish Athens as a major center of learning in the West.

WIDE INFLUENCE: Engraving of the Muslim Jabir Ibn Hayyan (721–815) teaching the works of Aristotle to a class at the School of Edessa, in modern-day Turkey.

Some of Aristotle's ideas lasted even longer. Future generations of Greek and Roman thinkers employed his methods of logic. During the Islamic Golden Age (ca. 750–1250), he was revered as the greatest Western thinker and often called the "first teacher." Much philosophy from the Middle Ages (ca. 400–1400) worked on incremental improvements to Aristotle's work, and even after the Age of Enlightenment, which turned philosophy on its head, his work continued to influence notable thinkers.

Aristotelian logic provided a platform from which physics and the other sciences could be advanced. The principle of reaching conclusions via empirical premises that are testable and verifiable is the very foundation of all experimentation. By modern standards, Aristotelian logic is a flawed method; there are many areas in which it can be broken. It is also overly reliant on term logic, which, while nevertheless very strong, is limited in its use. Despite these issues, Aristotle is still seen as one of the greatest philosophers of all time and is often referred to as the "father of logic." Without his contributions, science simply wouldn't be what it is today.

ARCHIMEDES PROBABLY DOESN'T TAKE A BATH

Many people have heard about Archimedes (ca. 287–212 BC), though probably not for the right reasons. While he may not have run through the streets naked shouting "Eureka," he was an incredibly important and influential figure of early physics, whose major achievements were a product of the simple application of mathematics.

Not much is known about Archimedes' personal life. He was born in Sicily, and his father, Phidias, was an astronomer who may have been some relation to the king of the Sicilian province of Syracuse. Much more, however, is known about him through his works and his contribution to science.

Archimedes is best known for his part in the likely apocryphal tale of the king and the crown, in which a king in the mid-3rd century BC gave a jeweler a measure of gold with which to make a crown. Upon being presented with the crown, the king suspected that the jeweler had mixed in silver so as to cheat the king and keep some gold. Naturally, the king wanted to verify this without damaging the crown. He called for the great mathematician Archimedes. Weeks later, after another stressful day of getting no closer to figuring it out, Archimedes had a servant run a bath. The bath was filled to the brim, and when Archimedes got into the bath, water flowed over the side. He understood that the volume of displaced water was equivalent to the volume of his body and quickly realized that he could use a similar method to calculate the volume of the crown. Once he had the crown's volume, he would be able to check its density, and in turn check whether it was 100% gold. (Gold has a higher density than silver, so if silver had been mixed in then the density of the crown would be lower.)

APOCRYPHAL TALE: A 16th-century engraving of Archimedes—evidence that a memorable story can do much to cement a scientist's place in history.

Ecstatic with his breakthrough, Archimedes is said to have leaped from the bath and run naked into the street shouting "Eureka!" meaning "I have found it!" The story concludes with the demonstration that silver had been mixed into the crown, whereby the jeweler was swiftly executed, proving that not all tales have happy endings. The likelihood of this tale being true is slim; Archimedes himself never mentioned it in his own works, and the precision required to measure the volume accurately enough to calculate the density change simply wasn't feasible at the time.

ARCHIMEDES' PRINCIPLE

Archimedes did, however, study fluid displacement, and in his work *On Floating*

Bodies he set out what has become known as Archimedes' principle:

> *Any object, wholly or partially immersed in a fluid, is buoyed up by a force equal to the weight of the fluid displaced by the object.*

Another way of expressing this is that an object in a fluid (such as water) will seemingly lose weight equal to the weight of the fluid that is displaced. Its effects are discernible when you float in water, or when you are able to pick up normally heavy objects with ease from the bottom of a swimming pool. With this insight, Archimedes then did something very impressive—he created a formula to explain the effect. A modern-day version would be as follows:

$$W_o + W_d = W_a$$

Where W_o is the original weight of the object, W_d is the weight of the displaced fluid, and W_a is the apparent weight of the object in the fluid. It is possible that he would have been able to solve the crown problem using this formula. If identical water-filled containers were placed on each arm of a set of scales, with the crown submerged in one container and an equivalent weight of gold in the other, a difference in density between the two objects would produce different apparent weights. This is to say that if the crown were not solid gold, the scales would tip.

Archimedes was one of the first to use mathematics to explain physical phenomena. As you are likely aware, physics is now seemingly overrun with equations that describe almost everything in exquisite detail. As we've seen here, the technique was first used by Archimedes. What's more, he also invented the field of hydrostatics (the study of nonmoving fluids).

THE GREAT INVENTOR

Archimedes' other achievements include discovering infinite series (the precursor to modern calculus), calculating the volumes of spheres and cylinders, and describing parabolas. He is also famous for his many inventions.

He created the Archimedes Screw, which consists of a screw inside a pipe. When turned, it allows liquids to be transported upward through the pipe. The benefits of the invention quickly became clear, and it was adopted across the ancient world for irrigation and the draining of mines or fields. It could potentially even have been used to transport water throughout the Hanging Gardens of Babylon. The basic concept is still being used today in combine harvesters, water treatment plants, and even chocolate fountains.

Many of Archimedes' most interesting inventions were designed to defend the city of Syracuse against the Republic of Rome during the Second Punic War (218–201 BC). The siege of Syracuse was focused around the naval blockade of the city's supply lines. Archimedes had already worked on improving the accuracy and effectiveness of the city's catapults and ballista, but he realized that new technologies were needed.

His first invention was the Archimedes' Claw. Shaped like a trebuchet, it would drop a grappling hook onto an enemy ship before its heavy load arm was allowed to swing

MOVING THE WORLD

In addition to his work on fluid displacement, Archimedes also studied the mathematical principles of levers. It was long known that levers make it possible for humans to move objects which otherwise would be too heavy for them to shift. Archimedes presented his findings in a formula, a modern-day version of which is:

$$F \times D = T$$

Where F is the force applied to the lever, D is the distance from the point where the force is applied to the fulcrum (the balancing point of the lever), and T is the resulting torque (a rotating force) around the fulcrum.

The formula shows that by increasing the distance from which you apply the force you can increase the resulting effect. This is why wheelbarrows have the wheel at the front, and door handles are placed on the opposite side to the hinges. The effect of this is both universal and so potent that Archimedes is quoted as saying, "give me a place to stand and I shall move Earth." In other words, if he had a lever long enough and a place in which to push it, he would be able to lift Earth.

This is true. However, a quick calculation shows just how far away that place would be. With the weight of Earth as 5.972×10^{24}kg and the amount of pushing force a human can apply as 675 Newtons, Archimedes would have needed to stand 8.68×10^{22}m away—about 9.2 million light-years away! That is just over three and a half times farther than the nearest galaxy, Andromeda.

downward, thus, in theory at least, pulling the boat upward. The hoped for result was that the crew would be dislodged, or that the vessel would be damaged or sunk. Another invention, misleadingly called Death Ray, consisted of an array of mirrors arranged in such a way that they would focus and redirect the sun's light onto an enemy ship, which was reportedly set alight as a consequence. However, modern attempts to replicate this have failed. The device was more likely used to dazzle sailors, obscuring their vision of the coastline and making them less effective in combat.

In spite of Archimedes' efforts, the Romans eventually broke through and, despite orders to the contrary, he was slain in his home. The city and keep were overrun before being looted and razed to the ground.

PTOLEMY PLACES US AT THE CENTER OF THE UNIVERSE

During Ptolemy's lifetime (ca. AD 100–170), it was for many a given that Earth was at the center of the universe. Ptolemy, however, was not prepared to simply accept what he was told, and he sought to find mathematical proof. In so doing, he created a system of prediction that lasted for over 1,000 years.

For an important historical figure whose ideas helped shape early science, little is known about Ptolemy. It is likely that he lived in Egypt under Roman rule, and from his writings and his use of Babylonian data it has been concluded that he was probably of Greek descent, though it has also been claimed that he was of Egyptian origin, possibly even coming from royalty.

Ptolemy was not the first to believe that Earth was the center of the universe. In fact many, if not most, of the great thinkers that preceded Ptolemy, including Aristotle, believed that our universe was geocentric ("geo" derives from the Greek for "Earth"). It was a common trope of mythologies and religious texts from many parts of the world, not just Greece. Ptolemy consolidated the knowledge, ideas, and information of centuries of work by Babylonian and Greek scholars into a formalized model. This model didn't just state that Earth was the center of it all; it used geometry and calculations to prove it.

Ptolemy's works, such as *Almagest*, are among only a few works from that time to have survived. This has increased their importance as primary sources of information about early Greek trigonometry and its creators, such as Hipparchus, who contributed much to mathematics. Despite their importance, many of Ptolemy's works disappeared in the middle of the Dark Ages, resurfacing in approximately the 12th century, when they were rediscovered and translated into Latin.

WHY THE GEOCENTRIC MODEL?

A casual observer can be forgiven for thinking that everything orbits around Earth. It is plain to see that the sun, stars, and planets move across the sky. Standing on its surface, it is not possible to discern the rotation of Earth, so it is only reasonable to conclude that it must be the other things that move. In the 2nd century AD, Earth was known to be spherical, and it was a reasonable assumption that all the other heavenly bodies were too. From this and the observation that the bodies move across the sky and seemingly around us, it would have been natural to conclude that everything orbits around Earth. In support of this we see the universe as unchanging; year after year, the stars remain in the same

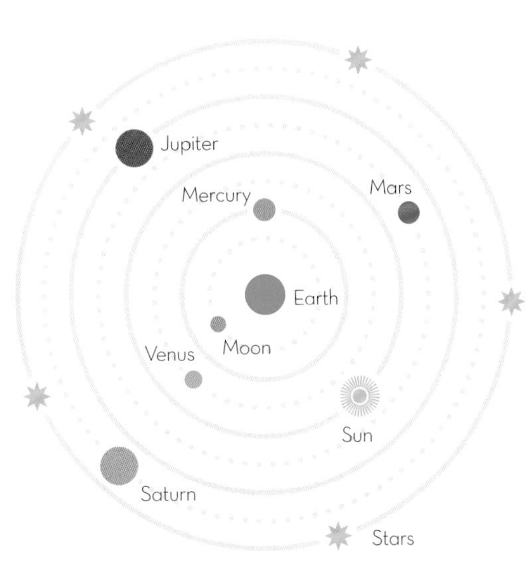

HEAVENLY SPHERES: A diagram of the heavenly spheres that Ptolemy set out in his book *Almagest*. Note that all of the stars sit on a single "celestial" sphere.

patterns, holding the same place in the star charts. It would be expected that if we were the one moving, the positions of the stars would change.

There was another, higher justification for this geocentric model: the gods, sitting upon their thrones atop Mount Olympus. It was only natural that the whole universe would circle around them and hence around Earth. This religious explanation was one of the reasons that the geocentric model lasted for so long. It was adopted by a number of religions, most notably various branches of Christianity, and it became a key symbol of the importance of man in the eyes of God. Anyone who challenged this view of the universe faced vehement opposition, as was the case with Copernicus (see pages 36–39) and Galileo (see pages 48–50).

It should also be remembered that the Ptolemaic system worked. Right through to the Middle Ages, Ptolemy's model was used to accurately predict the motion and positions of the heavens. While Ptolemy was alive, alternative systems were proposed (some of which placed the sun at the center), but his model performed better. It wasn't until much later that serious problems with the system became evident.

PTOLEMY'S GEOCENTRIC MODEL

Ptolemy lay out his model in *Almagest* (also known as *Syntaxis Mathematica)* a series of thirteen books that address everything from the motions of each planet to the length of days. As set out in the first book, Ptolemy makes five major claims:

1 Earth is a sphere
2 Earth is the center of the universe
3 Earth does not move
4 The stars exist on a celestial sphere that moves as a solid sphere
5 Earth is very small compared to the celestial sphere and must be treated as a point

From here Ptolemy laid out his universe, placing the objects on circular orbits around Earth in the following order:

1 The moon
2 Mercury
3 Venus
4 The sun
5 Mars
6 Jupiter
7 Saturn
8 The celestial sphere (all the stars)

This, however, isn't quite the complete picture. Just having circular orbits quickly throws off predictions, making them useless.

10
20
30
40
50
60
70

Ptolemy referred to the original circular orbit as the "deferent" and stated that the planets move in a smaller orbit around this larger orbit called an "epicycle." This had been done before by Hipparchus (the inventor of trigonometry, on whose work Ptolemy based much of his own), but it only worked for Mercury and Venus.

Ptolemy's great contribution was the addition of "eccentricity." He postulated that, while the model was geocentric, the planets' orbits were not directly centered on Earth but rather at some point slightly away from it. This accounted for deviations in the movement of the other planets. Both Ptolemy and those who worked on geocentric models after him focused on using these "orbits within orbits" to explain various events and discrepancies in what they saw.

Almagest also includes a detailed star catalog of 1,022 stars that sets out 48 constellations, which became the standard in Greece and were later included into the 88 modern constellations that are used today. This star catalog may have been copied from Hipparchus's works, along with the constellations, though whether this is true is unclear.

GEOCENTRIC ISSUES

Despite the model's prominence, it was not accepted by everyone. Some thinkers disagreed with much of his mathematics and rallied around the heliocentric system proposed by Aristarchus of Samos nearly 400 years earlier. The Ptolemaic system, with its many rings and multiple orbits, was complex.

Even some of the features of Ptolemy's geocentric system that compared favorably with the heliocentric model, such as its explanation of the apparent nonmovement of the stars, had already been explained by people such as Archimedes. He suggested that the stars were merely very far away, which would cause this effect and work with the geocentric model. One of the most obvious issues with Ptolemy's model was its inability to explain the procession of Mars. While many of the planets perform small circles in the night sky (which was explained by the epicycles), Mars does something a little different. It moves from west to east, but occasionally shifts downward in the night sky and then west, shifting downward once again before moving back east to rejoin its regular orbit. This S-shaped motion could not be explained by any number of epicycles and became a key proof of the heliocentric model.

But perhaps most damning of all was that by the 1400s and 1500s, the predictions were starting to fall apart, like a cheap watch losing seconds, then minutes, and now being hours out of time. Eclipses weren't happening when they were supposed to, and the Julian calendar wasn't accurate to the equinox. Sailors were forced to adopt new methods of positioning, creating the longitudinal system still in use today. The message was clear, Ptolemy had done very well, but it was time for a new system.

LEGACY: A portrait of Ptolemy from 1584. His teachings remained widely influential until the late 18th century.

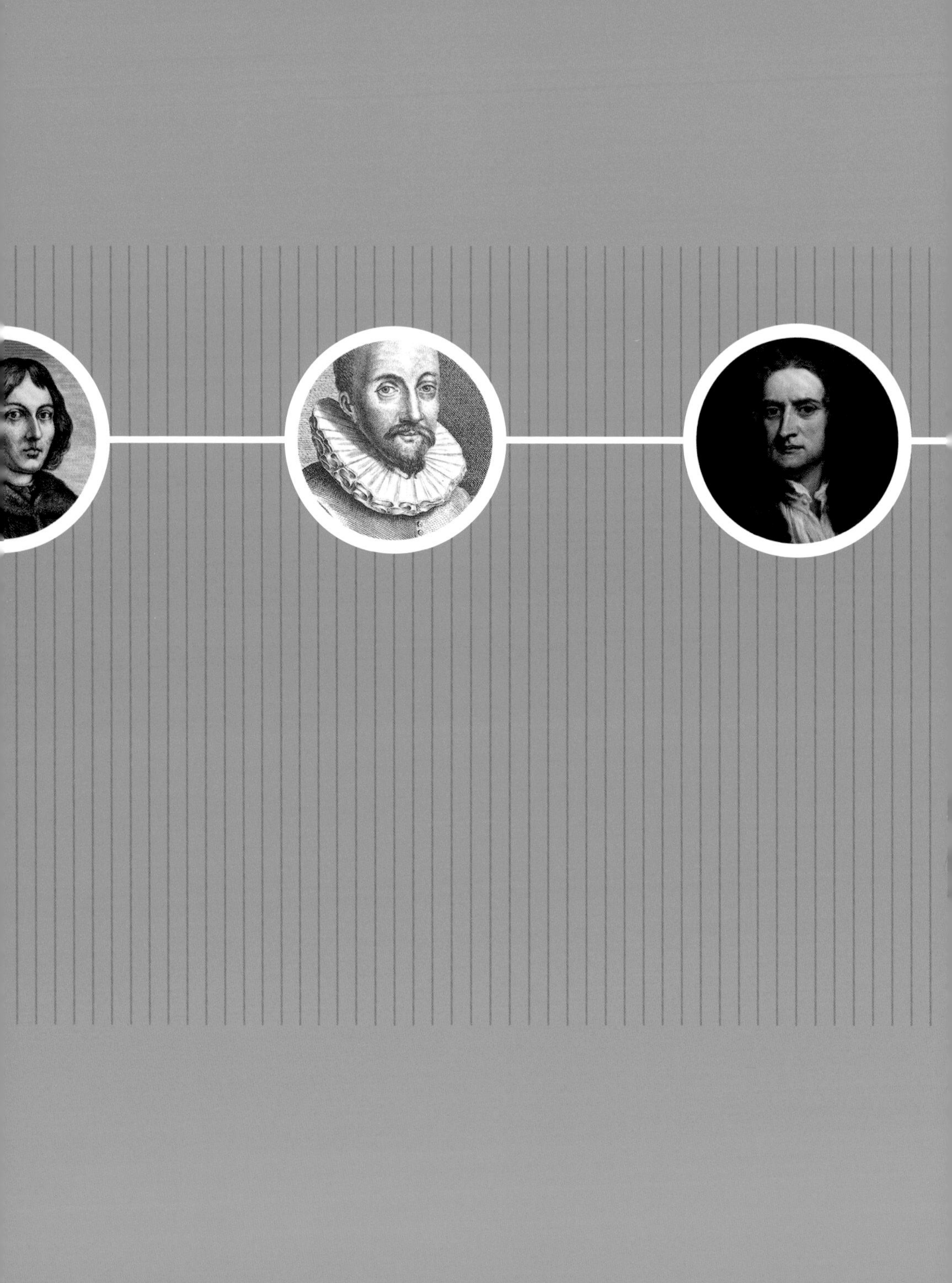

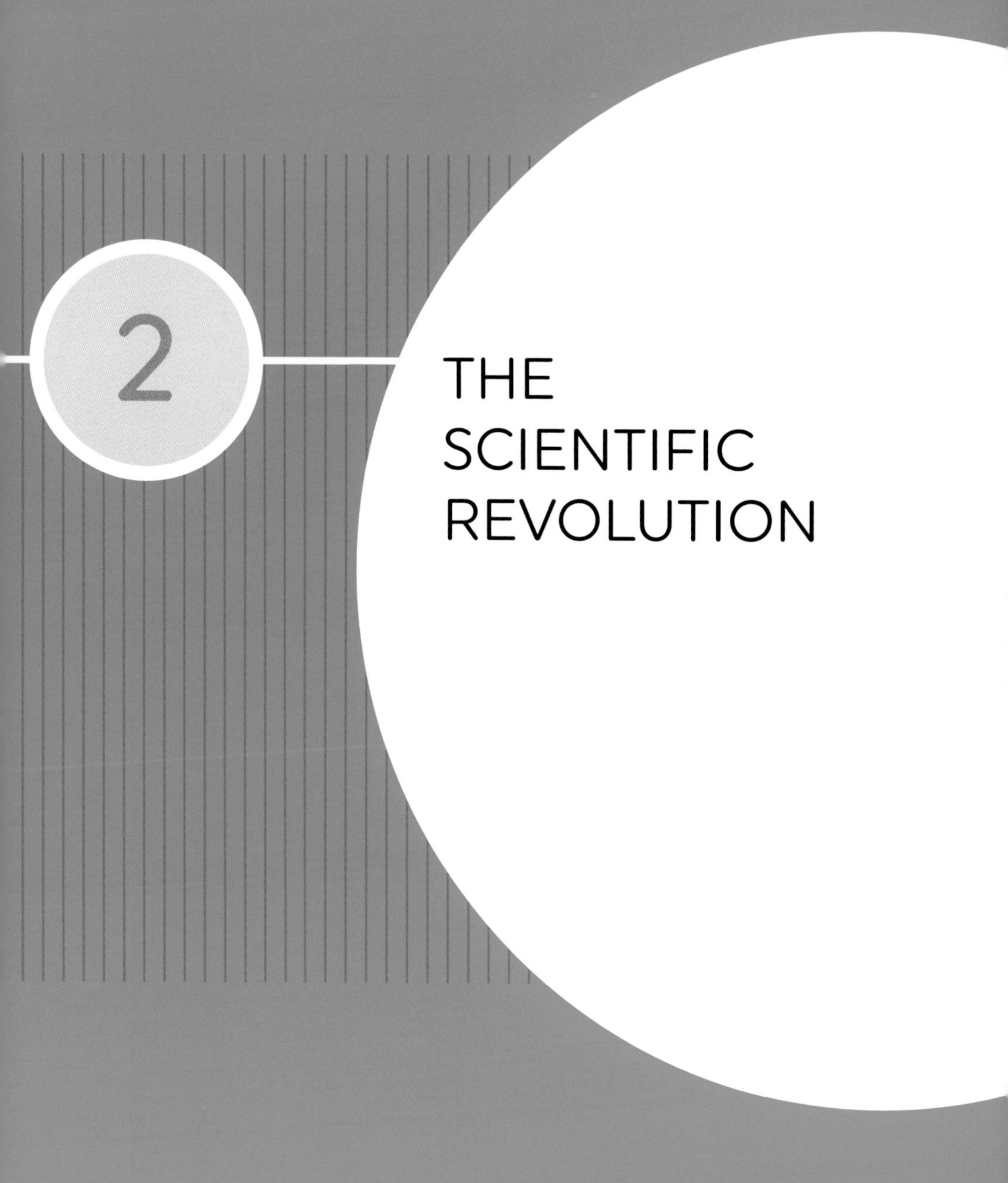

2 THE SCIENTIFIC REVOLUTION

IBN AL-HAYTHAM DESCRIBES LIGHT

Al-Hasan Ibn al-Haytham (965–1040) was the first scientist to accurately describe the properties of light and the principles of vision. He was also arguably the first theoretical physicist. His contributions helped begin the scientific revolution that would change the world.

Ibn al-Haytham was born in the city of Basra, in what is now Iraq. At some point in his early life he moved to Cairo, where he was aided by the Caliph's (the local leader and lawmaker) patronage of the sciences.

While he worked across a great many disciplines, including mathematics, geometry, and astronomy, Ibn al-Haytham's greatest work was the *Kitab al-Manazir* (*Book of Optics*). It was a seven-volume series that discussed light and how we perceive it. Much of the work is focused on how the eye is able to perceive light, but in the first and last book he correctly described light as occurring as parallel rays. While his exact explanations were often not correct, he did describe the concept of refracted rays of light (the change in light when it enters a different substance—the same effect that causes a spoon to appear to break when you place it in water) and also set out the idea that an object can both produce its own light and have the light from other objects reflected off it.

A MAD MAN?

Shortly after moving to Cairo, Ibn al-Haytham is said to have proposed a mechanical method of helping to regulate the yearly flooding caused by the River Nile. He was given permission by the Caliph to begin work on the project, which included constructing a dam at the same site as today's Aswan Dam. However, he quickly discovered that, with the technology and resources available to him, it would be impossible to complete. Fearing the potential consequences of this, it is said he pretended to go mad, whereby he was placed under house arrest until the death of the Caliph. It was during this period, when he was confined to his house, that he did most of his major work in physics.

EARLY PHYSICIST: A drawing of Ibn Al-Haytham, one of the greatest teachers of the golden age of Islam and perhaps the first theoretical physicist.

THE BEGINNING OF THEORY

Ibn al-Haytham was arguably the first to use the scientific method (see pages 54–57), which would take another 500 or so years to develop fully. He started with a solid mathematical theory, based on casual observation, and the work of his contemporaries, and then designed a number of experiments to test his theory. For example, in one experiment, during which he measured the intensity of light passing through a gradually shrinking hole, he didn't just take two measurements. He set up a systematic method of testing it at set points, and even possibly repeating the experiments to test for reliability. Both techniques are now standard practice in most experiments, but they were new at the time.

While Ibn al-Haytham's method was imperfect and didn't catch on immediately, the precision of his experimentation and the broadness of the topics he covered in his 45 published works earned him much respect, and his ideas on how science should be done slowly started to gain popularity. His methods spread across the world, leading thinkers such as Galileo and Sir Isaac Newton to adopt and develop the scientific method. For the importance of his work until the mid-1500s, al-Haytham was often called the "second Ptolemy" or "the physicist."

LADISLAUS THE POSTHUMOUS SENDS A LETTER USING ARABIC NUMBERS

Ladislaus the Posthumous (1440–1457) is a little-known European monarch. While his reign as a monarch was full of incident (with civil war and succession challenges), his relevance to the history of physics comes from his use of Arabic numbers.

Ladislaus the Posthumous was, by the end of his short life, the Duke of Austria and King of Bohemia, Croatia, and Hungary. Upon his birth in 1440 he was crowned king by use of the Holy Crown of Hungary, but the Diet of Hungary (the governing body of the country) declared the coronation invalid and instead offered the throne to Vladislaus III of Poland. Civil war broke out, and Ladislaus was moved into the court of Austria, under his guardian, Frederick III, King of the Romans. It was here that Ladislaus grew up and received his education.

The Austrian court was a hive of intellectuals, the most important of whom were the followers of Leonardo of Pisa, better known as Fibonacci, who had introduced the West to Arabic numerals 200 years earlier. Ladislaus remained in Frederick's court until 1452, when he returned to Hungary and made short work of affirming his position as king.

In 1456, only a year before his death, he did something unprecedented. In letters and court documents, he began to use the Arabic

CHANGING FORMS: The evolution of Arabic numerals into those familiar to us today took place over the course of centuries.

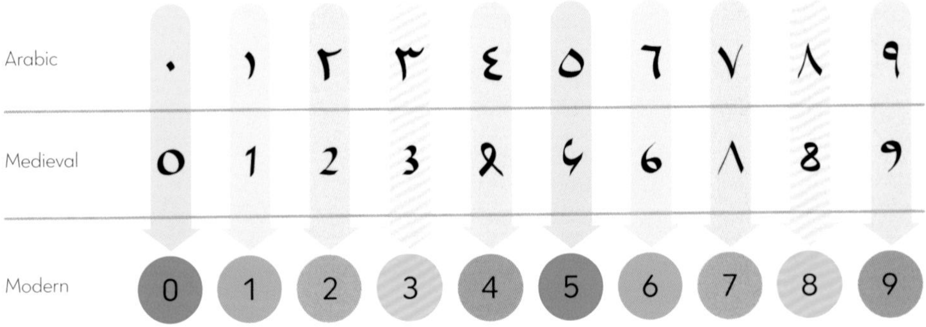

Arabic: ٠ ١ ٢ ٣ ٤ ٥ ٦ ٧ ٨ ٩

Medieval

Modern: 0 1 2 3 4 5 6 7 8 9

numerals in place of the more traditional Roman numerals. This was the first official use of them in the West, and it seemed Fibonacci's teachings were coming to fruition. From here their use spread across the Western world, and they eventually evolved into the modern numerals used today. This process was facilitated to a significant degree by printing presses, which used Arabic numbers in an effort to reduce costs, because the system required fewer symbols. By the mid-16th century the number notation that we know today was common in most of Europe, and it took only a few hundred years more before it was adopted the world over, in places such as China, Indonesia, and Russia.

WHY ARABIC NUMERALS?

Before Arabic numerals were adopted, most mathematics and physics was done using either Roman or Greek numbers. While these systems are perfectly functional, they can be difficult to get to grips with. They also take longer to read and are easier to make mistakes with. This issue lies with the fact that they use a compound numbering system. To illustrate this, we can look at the number 3,807 in Roman and Greek numerals:

Roman MMMDCCCVII
Greek XXX𐅅HHHΓII

These numbers are ascertained by adding all of the digits together. We shall work off the Roman numerals as they're more commonly understood, but the Greek numbers work in exactly the same way.

M represents 1,000. Because there are three of them, we get the number 3,000. This is followed by D, which represents 500, giving us a running total of 3,500. C is 100, and three of those added on gives us 3,800. V is 5, and I is 1, which gives a further 7, making the final total 3,807.

It should be noted, of course, that people at this time were brought up and introduced to mathematics through these compound numerals, and they would have found them much easier to use than you or I. Even so, it's easy to see why Arabic numerals were quickly adopted. Even at a quick glance, 472 × 76 is much easier to read than CDLXXII × LXXVI.

THE GREEK ALPHABET IN PHYSICS

Greek numerical symbols are still around and are used heavily in physics. Many are used to represent variables, such as ρ (rho, the seventeenth letter of the Greek alphabet) for density and Ω (omega, the twenty-fourth letter) for radial velocity. Others, such as π (pi, the sixteenth letter, and also the mathematical constant 3.14159...) or ϵ (epsilon, the fifth letter, and the mathematical constant 2.71828...) are as much numbers to a physicist as 6 or 42, and have remained mostly unchanged since their discovery by the ancient Greeks.

NICOLAUS COPERNICUS PUBLISHES *DE REVOLUTIONIBUS*

The Ptolemaic system of a geocentric universe was no longer working. It couldn't predict the goings on of the heavens anymore. Something new was needed, and a brilliant Prussian mathematician provided it. Even though he knew his work was controversial and revolutionary, he could not have predicted the effects it would have on science.

Nicolaus Copernicus (1473–1543) was a German-born polymath who worked as a doctor, governor, diplomat, and economist, during which time he created the quantity theory of money and Gresham's Law, both key economic concepts. He spoke Latin, German, Polish, Greek, and Italian, and worked for some time as a translator. It is clear to see that he was an incredibly talented man.

Copernicus may have started work on his heliocentric (*helios* is Greek for "sun") model as early as 1503, while working as secretary to his uncle in the Royal Prussian Diet (the Prussian government). By 1514 he had written a 40-page outline of his theory called *Commentariolus*, or "Little Commentary." It was a precursor to *De Revolutionibus* and in it he offered seven postulates (see box, page 38).

He circulated copies to his friends and select peers, but it was not published during his life. Copernicus was reluctant to publish any works that offered his heliocentric view, and it was easy to see why; huge pressure was exerted by leading scholars and religious interests in an attempt to protect the geocentric model. When rumors of the work circulated in 1539, the famous theologian Martin Luther said:

> *People gave ear to an upstart astrologer who strove to show that Earth revolves, not the heavens or the firmament, the sun and the moon… This fool wishes to reverse the entire science of astronomy; but sacred Scripture tells us that Joshua commanded the sun to stand still, and not Earth* [Joshua 10:13].

The work itself consisted of six books, laid out in a similar manner to Ptolemy's *Almagest*, which covered the moon, planets, and stars, and the mathematics that governs them. It was completed by around 1532.

A LIFE'S WORK

One of Copernicus's peers, Johann Albrecht Widmannstetter (1506–1557), a scholar of law and theology, gave a series of lectures

CHALLENGING DOGMAS: A drawing of Nicolaus Copernicus. He remained reluctant to publish his ideas his entire life, as he knew they would be controversial.

based on *Commentariolus*, which were attended by senior figures in the Catholic Church, including the Pope. The lectures generated much interest in the heliocentric theory, and a number of attendees subsequently wrote to Copernicus urging him to publish. He eventually instructed a friend in 1543 to have *De Revolutionibus Orbium Coelestium* (*On the Revolutions of Heavenly Spheres*) published.

Still worried about publishing such a controversial work, Copernicus began it with the following preface to the then pontiff, Pope Paul III:

> *I can readily imagine, Holy Father, that as soon as some people hear that in this volume, which I have written about the revolutions of the spheres of the universe, I ascribe certain motions to the terrestrial globe, they will shout that I must be immediately repudiated together with this belief. For I am not so enamored of my own opinions that I disregard what others may think of them. I am aware that a philosopher's ideas are not subject to the judgment of ordinary persons, because it is his endeavor to seek the truth in all things, to the extent permitted to human reason by God. Yet I hold that completely erroneous views should be shunned. Those who know that the consensus of many centuries has sanctioned the conception that Earth remains at rest in the middle of the heaven as its center would, I reflected, regard it as an insane pronouncement if I made the opposite assertion that Earth moves.*

Copernicus collapsed from internal bleeding or a heart attack in late 1542 and died on May 24, 1543. It is said that on the day of his death he was presented with the first printed copy of his work and allowed himself to die happy, knowing his life's work was complete.

The immediate reception was cool, and the initial print run of 400 copies did not sell out. This was likely due to the complexity of the book, which would have been prohibitive to many readers. However, many of those that could read it quickly took it up as the successor to *Almagest* and began pushing for it to be the accepted scientific theory.

THE SEVEN POSTULATES OF *COMMENTARIOLUS*

1. Celestial bodies do not orbit around a single point.
2. The moon orbits around the center of Earth.
3. All known objects rotate (rather than orbit) around the sun, which is near the center of the universe.
4. The distance between Earth and the sun is insignificant in comparison with the distance between Earth and the stars; hence there is no visible motion.
5. The motion of the stars is caused by the rotation of Earth.
6. The Earth is a sphere that moves around the sun, but it has more than one motion, which causes the annual migration of the sun.
7. Earth's orbital motion causes the irregularities observed in the movements of the other planets, such as apparent reverse motion.

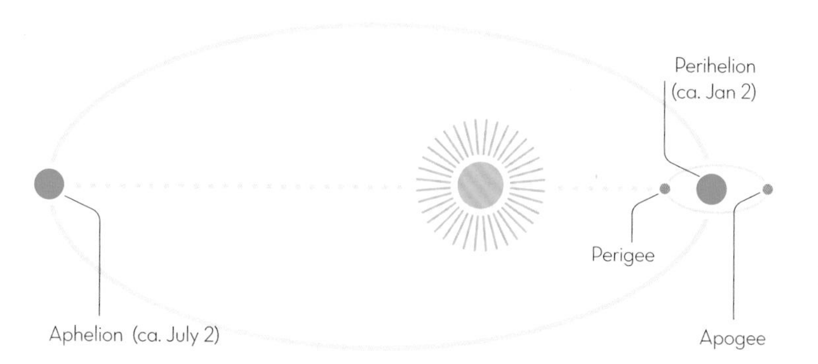

EARTH'S ORBIT: Earth's farthest point from the sun is the aphelion, its nearest point perihelion. The moon's orbit has an apogee (the farthest point from Earth) and perigee (nearest).

Despite Pope Paul III's interest, and Copernicus's attempts to avoid conflict with the church, it was soon under fire from a number of religious figures, and there were calls for severe measures to be taken against the work and its followers. In 1616 *De Revolutionibus*, along with many of the books that called for its acceptance, was placed on the Index of Forbidden Books by the Catholic Church and would remain there until 1758. Despite the Church's efforts to suppress them, the ideas became increasingly influential, so much so that they prompted a revolution.

THE COPERNICAN REVOLUTION

Copernicus's model had many critics, some of whom raised scientifically valid objections. But he had established a new line of thought, and others were keen to follow. One of the greatest contributors to this revolution was Johannes Kepler (1571–1630), a German scientist who became an assistant to the prominent Danish scientist Tycho Brahe (see page 40). Kepler was one of the few scientists of his age to openly support Copernicus's ideas. He continued the work of Copernicus, creating his eponymous laws of planetary motion:

1. Planets move in elliptical orbits, with the sun as one of the foci.
2. A line connecting a planet to the sun will always sweep out an equal area over an equal time.
3. The period of orbit around the sun is proportional to the point of farthest distance from the sun to $3/2$ the power.

This discarding of the traditional circular orbits in favor of the elliptical ones was a huge step and solved almost all of the issues that still plagued the heliocentric model. Galileo also contributed two major discoveries to the heliocentric picture of the solar system in 1610: the moons of Jupiter and the phases of Venus.

It is generally accepted that this period of significant scientific advance, known as the Scientific Revolution, concluded with Isaac Newton. His work *Principia* (see pages 54–57) gave a solid mathematical explanation for how the planets moved and why that matched up with the heliocentric model and Kepler's laws of motion. It worked so well, in fact, that it would remain unchallenged for over 200 years, until Einstein presented his theory of general relativity (see pages 124–127).

A NEW STAR APPEARS IN THE SKY

Even as the concept of a geocentric universe was being challenged, the heavens were still thought of as unchanging. But in 1572 a bright new star appeared in the constellation of Cassiopeia, and once again the foundations of astronomy were shaken.

In early November of 1572, in the constellation of Cassiopeia, a new star appeared that was comparable in size and luminosity to the planet Venus. From China to England, monarchs summoned the greatest thinkers of the age to interpret this new development. What was it and what did it mean? The new star gradually faded and by 1574 had disappeared from the sky, but its legacy continues to be felt.

The event was witnessed by a great many people, one of the most important among them the Danish astronomer Tycho Brahe (1546–1601). He was a man of exacting scientific methods. Over his lifetime he built a great number of astronomical instruments and even a research institution, where he was able to produce observations many times more accurate than any others of the day. He believed in mathematics above all else, and he was supportive of Copernicus, despite continuing to endorse the geocentric model of the universe. As a result, he created his own system, which incorporated the geo- and heliocentric viewpoints.

PARALLAX

Even before the appearance of the new star, Tycho Brahe was skeptical of the view that the stars in the sky were fixed and unchanging, and he predicted that there should be a small parallax in the stars over a 6-month period. Parallax is the illusory movement of a static object, which is in fact a product of movement of the point of observation. It is easy to achieve yourself: Simply hold this book open in front of your face with one hand and close one eye. With the other hand, position an outstretched finger between yourself and the book, and align it vertically with the spine. Now switch your closed and open eyes; you will find that the finger seems to have shifted, despite not having actually moved. This illusory movement, which is known as parallax, does exist in the stars, but it is so small that it couldn't be observed until the early 1800s. Brahe pointed to this lack of discernible shift as evidence that the stars must be very far away, at least 700 times the distance between Saturn and the sun (the closest star is actually about 29,000 times the distance from Saturn to the sun, but he was technically correct). This also meant that they had to be very large.

COLORFUL PAST: A painting of Tycho Brahe from 1596. He wore a fake nose after losing his real one in a duel with another student at university.

EFFIGIES TYCHONIS BRAHE OTTONI DAN:
ÆTATIS SVÆ ANNO 50 COMPLETO.
QVO POST DIVTINVM IN PATRIA
EXILIVM LIBERTATI DESIDERATÆ
DIVINO PROVISV
RESTITVTVS EST.

About a year after this new star appeared in the sky, Brahe published *De Nova Stella* (*The New Star*), which gave this phenomenon its name, a supernova. In it he stated that, as this new object showed no parallax, it must be at least farther away than the moon. Furthermore, as it remained in a fixed place relative to the stars for such a long period, it must also be farther away than the planets, and was most likely part of the celestial globe. This was a huge discovery. The Aristotelian world was founded on the idea that the heavens were unchangeable, and yet here was proof that they could change. This enhanced the credibility of the Copernican system and therefore undermined long-held beliefs regarding the universe.

Although the rough position of the supernova, labeled SN 1572, had been known for a long time, in 1952 astronomers at the Jodrell Bank Observatory in England were able to accurately measure its location. Over the following years, it was observed by many telescopes, and its remnants were revealed as an incredibly beautiful cloud of gas exploding outward into space.

WHAT IS A SUPERNOVA?

Stars are mostly made from hydrogen. As they burn, hydrogen particles fuse together to create helium, a reaction that produces large amounts of energy. Eventually the stars begin to run out of hydrogen, and the pressure created by this reaction decreases, at which point the star begins to collapse inward. If a star is large enough, it will, as it collapses, produce enough heat that the helium particles fuse together. This too will eventually run out, and it begins to collapse once more. Large-enough stars may continue to move up the periodic table, fusing different elements, but even the largest star will eventually stop at iron.

While the reasons for this are a little complicated, in essence this happens because the fusion of iron doesn't produce energy, but

MOVING STARS? This image shows how, at different points in the year, the stars can appear to move. In January, star B would be invisible behind star A; however, in July it would be easy to see, but star C would be behind star A. None of the stars have really moved, but how we perceive them has.

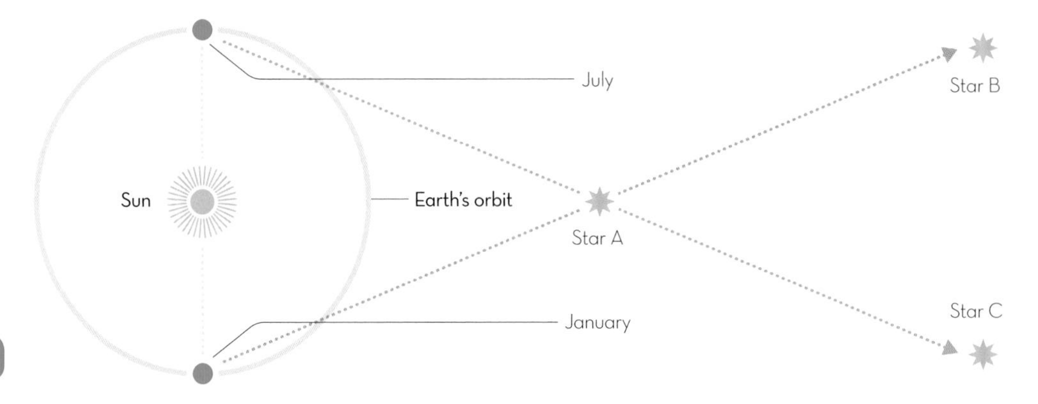

CHARTING THE HEAVENS

The realization that stars were perhaps not as static as once thought spurred many into believing that better catalogs of the stars were required. One of the most widely used was created by Tycho Brahe. In his typically thorough style, the catalog was incredibly accurate and contained a huge quantity of data. Over the years a great number of star catalogs have been created, such as the *Histoire Céleste Française* (*French Celestial History*; 1801), which cataloged 47,390 stars. Some catalogs are more specialized, looking at nonstellar objects (those that resemble, but are not in fact, stars), such as the Messier catalog of nebulae and star clusters (1784). Today one of the most complete catalogs is the SIMBAD Astronomical Database, which contains information on 8,455,904 celestial objects (as of August 2016).

rather takes energy. So at this point the stars collapse entirely. What happens next depends on the size of the star. If the star has less than about eight times the mass of the sun, it will become a white dwarf. This is where the collapse of the star is halted because there's no room to collapse inward, and the star becomes a small, hot ball of plasma that very slowly cools off over billions of years. If, however, the star has between eight and 50 times the mass of the sun, the result is a little more interesting.

As the star collapses rapidly, the mass of its core exceeds something called the Chandrasekhar limit—about 1.4 times the mass of the sun. When this happens, the core implodes. The rest of the star then collapses into this new space at speeds reaching a quarter of the speed of light, which causes the temperature to rocket to billions of degrees. This extreme temperature then causes the neutrons in the star to degenerate, and this causes a tremendous explosion. This explosion is of around 10^{44} joules every single second, which is equivalent to about 6.3×10^{31} nuclear bombs.

SN 1572 was a little different, however, because it was a type Ia supernova. It died and became a white dwarf rather than a supernova. However, it was orbiting another star and began leeching material off it. It dragged stellar gases and plasma across space from its companion star and onto its own surface. This continued happening until it gained enough mass to reach the Chandrasekhar Limit of 1.4 solar masses. At this point the core collapsed and it went supernova.

Until around 100 years ago, about eight or so supernovas had been documented, but in the last century, with the advent of space telescopes, we now discover them at an incredible rate, and find more each month than we had in the preceding several thousand years.

HANS LIPPERSHEY DESIGNS THE FIRST PRACTICAL TELESCOPE

When you stop and think about it, you realize just how incredible it is that until the early 17th century every astronomer, from Plato and Aristotle to Copernicus and Brahe, had only eyes to view the heavens with. Hans Lippershey's (1570–1619) invention of the first practical telescope provided astronomers with a powerful new tool that dramatically broadened their scientific horizons.

The use of glass for seeing had been common practice since the Greeks in 750 BC, who manufactured glass domes to read letters, in the words of the ancient philosopher Seneca the Younger, "however small and dim." By the end of the 16th century, lenses and mirrors had become increasingly popular and the techniques to create them had grown increasingly sophisticated. There was a boom in the Italian trade of fine glassware, fueled by the Murano glassworks in Venice. Despite attempts to keep the secrets of glassmaking contained, it eventually spread across Italy and as far as the Netherlands.

It was in the Dutch city of Middelburg where Hans Lippershey set up a spectacle workshop in 1594. How exactly Lippershey came up with the idea isn't known for certain; it has been claimed that he stole the idea, or perhaps that it came from his apprentice. Another story attributes the initial discovery to two children, whom Lippershey observed playing with some old lenses. They had realized that, by holding two of them together, it would make the weather vane on a nearby church look much larger. Lippershey followed suit, mounting two lenses securely inside a tube so that they could be properly focused.

At this time, the Dutch were fighting for independence from Spain in the Eighty Years War (1568–1648). One piece of Dutch military equipment was "perspective glasses," which allowed commanders to see farther on the battlefield. The glasses, which had a single lens, were something of a precursor to the telescope. Lippershey wasted no time, applying for a patent in the same year as he had finalized his design and presenting the device to the military. While they quickly adapted the design into binoculars and used a rock crystal lens, which was more durable, Lippershey's basic telescope design spread across the world, becoming available in Paris by early the next year and in Germany by the following autumn.

THE FIRST TELESCOPE?

There are two other claims to the title of inventor of the telescope. The first belongs to James Metius, whose brother had studied under Tycho Brahe. He submitted a patent only a month after Lippershey and petitioned the government of Holland, saying that he had previously created a telescope as good as the one that had been presented to them by Lippershey, and that he would (with a little monetary encouragement) be able to produce even greater results. The petition was dismissed and Metius refused to show his work to anyone, ordering that all his tools be destroyed upon his death so that nobody could learn from his techniques. The other claimant is Hans Jansen, who claimed that his father had created a telescope in 1590. However, Jansen was a suspect man who made counterfeit currency and became a convicted forger. His sister did corroborate the story of their father building his own telescope, but dated the invention between 1611 and 1619. So the invention is normally credited to Lippershey. Due to the fact that "many other persons had a knowledge of the invention," Lippershey's patent was declined, which facilitated the rapid dissemination of the invention across the world.

HOW DOES THE TELESCOPE WORK?

Hans Lippershey's design was for a refracting telescope. Light from stars, or from any object sufficiently far away (typically farther than 20ft, or 6m), comes as parallel light rays. A refractive telescope uses a convex (meaning curved outward) primary lens to refract the light and bend it inward. The light rays then focus to a point and start to spread out again. Just beyond this focal point, a secondary lens (known as an eyepiece) refracts the light again, this time causing the diverging light to become parallel. It also serves to reduce the distance between the light rays and makes distant objects appear larger. This, however, has the effect of making the object appear upside down.

How good a telescope is depends on the quality and size of the lenses. Poor-quality

MAGNIFICATION: A diagram of how lenses in a telescope magnify objects. Note how the parallel lines on the eye side are closer together than those on the object side.

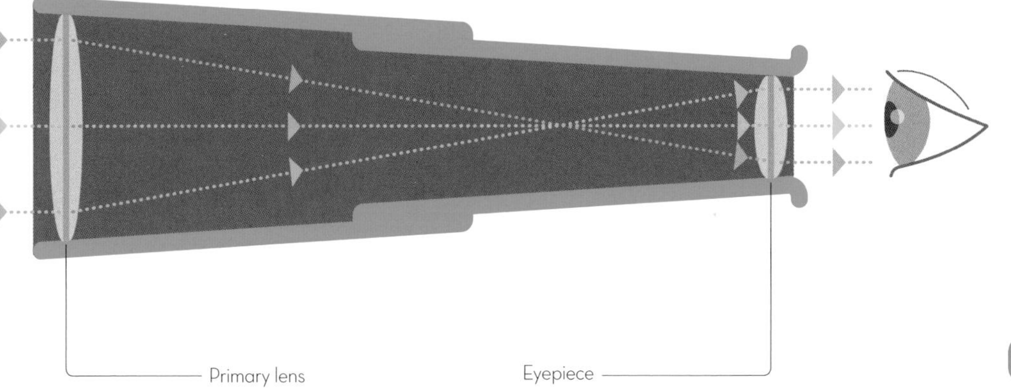

OTHER TYPES OF TELESCOPES

In addition to the refracting telescope, there are two other major types of telescope: the reflecting telescope, also called the Newtonian telescope; and the catadioptric telescope. Both of these use mirrors instead of lenses, which allows for much shorter focal lengths.

The Newtonian telescope uses a large primary mirror instead of a primary lens. The mirror collects the light, which is then focused onto a smaller secondary mirror and reflected onward to an eyepiece lens (or series of lenses), which converts it back into parallel light. This does, however, require that the eyepiece be placed facing off the side of the telescope, which can limit how observing is done.

The catadioptric telescope works in a similar way to the Newtonian, using a large primary mirror to focus the light onto a secondary mirror. However, in this set-up the light is reflected back through a hole in the center of the primary mirror and into the eyepiece lens. This allows the catadioptric telescopes to be more compact than Newtonian telescopes of equivalent power.

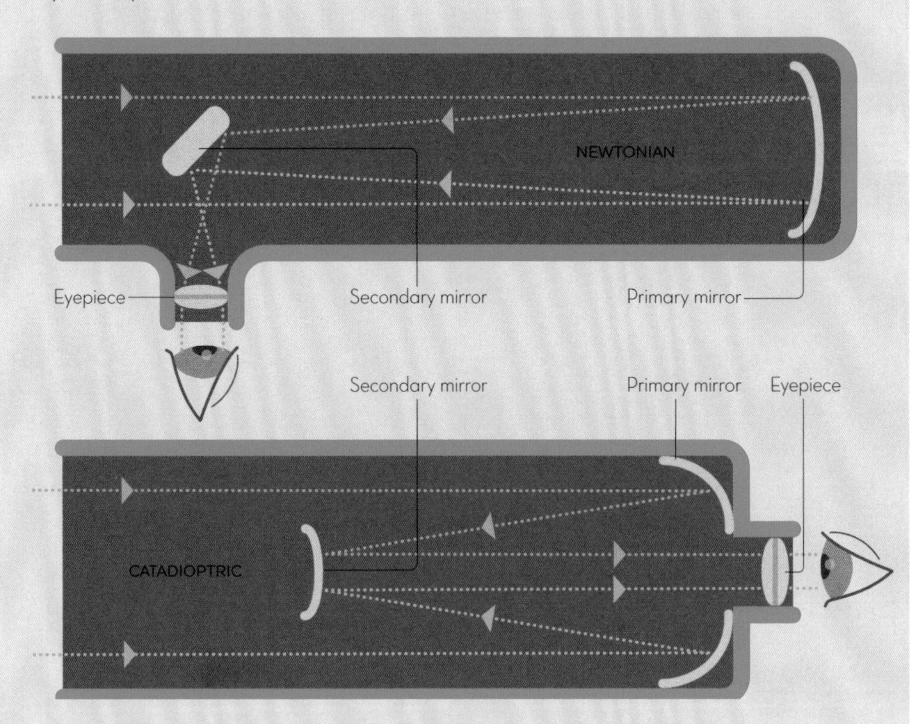

lenses might not refract the light correctly, leading to blurry images or incorrect coloring. The bigger the primary lens, the more light it is able to take in, thus increasing magnification. However, this then also requires a longer tube, because the distance to the focal point is longer. The largest refracting telescope ever built was for the World's Columbian Exposition in 1893, after which it was installed at the Yerkes Observatory in Wisconsin. The primary lens is 40in (100cm) across, and the telescope is 43ft (13m) long, in order to accommodate the long focal length.

MODEST RETURNS: A drawing of Hans Lippershey from 1655. Despite his attempts to profit from the invention of the telescope, he was left largely unpaid for his work.

THE USE OF TELESCOPES

Telescopes were quickly adopted by the astronomical community. Galileo was observing the moon of Jupiter by 1610 and people across the world were making discoveries at an incredible rate. Thousands of new stars, planets, moons, and even other galaxies became visible. Thanks to the telescope, it became possible to see distant phenomena such as stellar nebulae (the birthplaces of stars) and huge interstellar dust clouds.

Telescopes were used not only by astronomers; maritime and military personnel found uses for them, and they were even adopted as play things by the noblemen of the age. Galileo is quoted as saying: "Oh! When will there be an end put to the new observations and discoveries of the admirable instrument?"

In truth, there has never really been an end. Since their invention, telescopes have become larger and more precise. The limiting factor became not the availability of high-powered telescopes but rather the manpower and time needed to view ever smaller sections of the night sky. Today we have launched telescopes into space, where they are able to observe constantly unhindered by weather or our atmosphere. Despite our best attempts, and the use of computers to catalog and map the skies above us far faster than any human could ever hope to do, we have at a very generous guess managed to view perhaps one percent of the observable universe. As the technology improves, it will be even less than that.

GALILEO IS TRIED FOR HERESY

Galileo Galilei (1564-1642) is undoubtedly one of the greatest physicists (and scientists) of all time. His achievements and discoveries are many, and he directly contributed to the dawning of the Scientific Revolution. However, being such a revolutionary thinker often placed him at odds with the Catholic Church, which in 1633 decided to take action against him.

Galileo was born in 1564 in Pisa. His father was an important composer who studied vibrations and sought to understand how acoustics could be improved. It was likely his father's influence that gave Galileo his love of physics and, more importantly, his preference for experimentation over pure mathematics. On his father's urging, he began a medical degree at the University of Pisa, where he changed his degree to mathematics and natural philosophy. There, Galileo worked on pendulums and created a thermoscope, before taking on a teaching role. In 1589 he was appointed the chair of mathematics before moving to Padua in 1592, following his father's death.

STARRY MESSENGER

In around 1610 Galileo obtained a telescope, and with it he made two major discoveries. The first was that the surface of Venus had phases very much like those of the moon; the second was that Jupiter is orbited by four bright objects (what we now know as the Galilean moons). These were important because Venus's phases could only be explained by the heliocentric model, and if Jupiter is orbited by other objects, it would provide proof that not everything in the universe orbits Earth. He published his findings in *Sidereus Nuncius* (*Starry Messenger*), which proved controversial. However, many were forced to concede that his findings were correct after successfully replicating his results. Many people nonetheless doubted his findings on principle. In a letter to Johannes Kepler in late 1610, Galileo lamented that many of those who opposed his views had not even looked through a telescope.

In 1615 *Sidereus Nuncius* was presented to the Vatican's Roman Inquisition as a heretical work. Galileo heard of this and traveled to Rome to clear his name, despite the urging of his friends that he remain at home. On February 24, 1616, the verdict of the Inquisition was delivered, declaring that:

> *The idea that Earth revolves around the sun is foolish and absurd in philosophy, and formally heretical since it explicitly contradicts in many places the sense of Holy Scripture.*

UNCOMFORTABLE TRUTHS: A painting of Galileo from 1636. His personal relations with the Church saved him from serious persecution, but his ideas proved too radical for many contemporaries.

Galileo was ordered to abandon his belief that the sun is the center of the universe and withhold from teaching or defending the position. The Inquisition also ordered the banning of a great many works that argued in favor of the heliocentric model.

HOUSE ARREST

This might have been the end of the matter, but in 1623 Galileo published *Dialogue Concerning the Two Chief World Systems*, which was heavily critical of the geocentric model and painted its proponents as simpletons who ignored evidence. In 1633 Galileo was called to Rome to stand trial. The somewhat condescending tone of his book won him few friends within the Church, and in June of that year the judgment "suspect of heresy" was passed. Galileo was sentenced to imprisonment, the printing of his works was halted, and *Dialogue* was banned. The following day, the sentence of imprisonment was reduced to house arrest, probably by order of the pope.

The justification given for the decision was to "protect the faithful from the disturbing effect of an unproved hypothesis." The sentence was light and it was passed without fanfare. Galileo was not asked to repent and his faith was never questioned. But the Church's position on the heliocentric model was clear; it would fight against those who held it to be true. Tensions between the Catholic Church and the growing number of Protestant states were high, and the Vatican was under attack for being weak on heretics. Although there was some sympathy for Galileo's plight, the Church simply couldn't allow him to get away with questioning their interpretation of scripture.

Throughout the trial and after his conviction, Galileo remained a devout Catholic. Despite the conviction, he remained on good terms with much of the Vatican, even receiving a special blessing from the pope shortly before his death. While under house arrest, he continued his experiments and sent manuscripts to be printed elsewhere in the world. Despite the Church's attempts to quash Galileo's works, they spread and became increasingly influential, not only as focal points for dissension against the Church doctrine, but because of the power of their many bold new ideas.

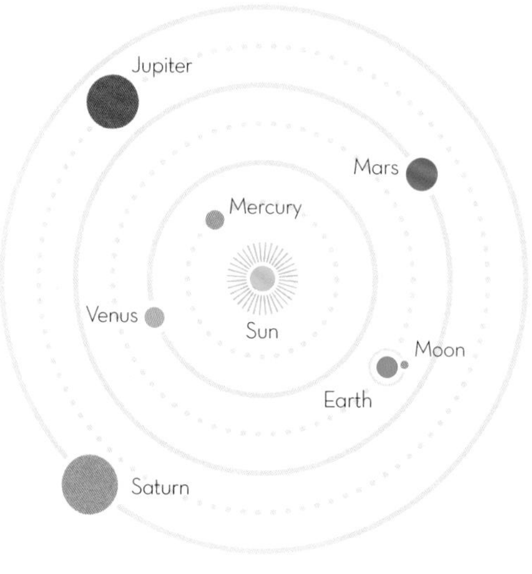

GUIDING STAR: The heliocentric model of the solar system placed the sun at its center, thus relegating Earth to the relative periphery. Despite opposition from the Church, the science could not be ignored.

THE FATHER OF MODERN SCIENCE

Galileo has been venerated and revered by almost every prominent scientist that succeeded him, from Isaac Newton to Stephen Hawking. Even the great Albert Einstein himself called Galileo "the father of modern science." Through his work, Galileo had established a new method, the experimental technique, and it would set the history of physics, and of science in general, on a new course. It was Galileo's greatest gift to the world.

It seems unthinkable today, but in Galileo's time, scientific theories could be accepted, and almost always were, based on mathematics and logic alone. So long as the math worked and it made sense, it would be accepted. Galileo believed that any theory should be backed up with experimentation, and his approach followed a number of steps:

1 Observe an effect and formulate an idea for how it works.
2 Create a mathematical proof that explains the phenomena.
3 Use this proof to predict an outcome of an experiment.
4 Perform the experiment to see if you get the intended result.
5 If the expected result is achieved, the theory is correct. If not, changes to the theory or experiment are required.

Most scientists of the day would have been happy to stop at step two, or just perform experiments. Galileo was the first to explicitly state that the laws of physics were mathematical. This combination of mathematics and experimentation was the final death knell for Aristotelian physics, and the experimental technique would be formalized and popularized by Isaac Newton, signaling the start of the Scientific Revolution.

AFTER GALILEO'S DEATH

Even at the time of the Galileo affair, the Church was far from absolute in its opposition and seemed not to reject entirely the science presented during the trial. St. Robert Bellarmine, a cardinal heavily involved in the trial, is quoted as saying that, were the heliocentric model to be proven, "it would be necessary to acknowledge that the passages in Scripture which appear to contradict this fact have been misunderstood."

After Galileo's death the controversy was largely forgotten. In 1718 the ban on republishing many of his works was lifted and by 1741 all of his works, including *Dialogue*, were available. This was followed by a gradual rolling back of the censorship targeting many of the books that supported the heliocentric model. By 1835, the Church no longer opposed it in any way. In 1992 Pope John Paul II declared that the Inquisition had acted in good faith but ultimately was incorrect. He also expressed regret for this mistreatment of Galileo and pledged to build a statue of him within the walls of the Vatican.

THE ROYAL SOCIETY IS FOUNDED

The Royal Society of London, England, is an incredibly important institution. Aimed at supporting and promoting scientific advancement through grants, training, and international cooperation, it has helped to set the standard for scientific research and continues to assist scientists.

How exactly the Royal Society began is unclear. It grew from a number of different groups, including the Gresham College group and The Philosophical Society of Oxford, which consisted of informal meetings of natural philosophers (physicists), whereby they would discuss their findings and work together to complete experiments and work out problems. Many of the groups adhered to the new scientific method, which, as seen in the previous entry, combined theory and experimentation.

After a lecture at Gresham College on November 28, 1660, a meeting of the College group decided that a "college for the promoting of physic-mathematical experimental learning" should be established. It was to hold a weekly meeting of discussion, in addition to running experiments. On July 15, 1662, King Charles II granted the organization a royal charter, transforming it into the Royal Society of London.

From here the society grew in importance and influence.

COPLEY MEDAL

In order to recognize the achievements of great scientists, the Royal Society created the Copley Medal for "outstanding achievements in research in any branch of science." It is still awarded today and stands as the longest-running scientific award in history.

It was first awarded in 1731 to Stephen Gray for "his new Electrical Experiments: as an encouragement to him for the readiness he has always shown in obliging the Society with his discoveries and improvements in this part of Natural Knowledge." He won it again the following year, for the rather less inspiring "Experiments he made for the year 1732." It has been awarded to many of the scientific greats, including John Goodricke, Steven Hawking, and Peter Higgs.

It developed into an organization that not only shaped modern science through its journal and inspired scientists to greatness with its awards and fellowships, but also reached out to engage the public through its Summer Science Exhibition, which is still hosted each year.

THE ROYAL SOCIETY: A late 19th-century wood engraving of a meeting of the Royal Society on Fleet Street. Isaac Newton is sitting in the president's chair.

PHILOSOPHICAL TRANSACTIONS

The Royal Society began publication of the world's first scientific journal under the title *Philosophical Transactions, Giving some Account of the Present Undertakings, Studies, and Labour of the Ingenious in many considerable parts of the World*, which was later reduced to simply *Philosophical Transactions of the Royal Society*. It began as a private venture of the then secretary of the society, Henry Oldenburg, who was to print it on the first Monday of each month, so long as he had enough to fill it. The first issue covered a number of topics, including recent improvements in glasses, the first account of the red spot on Jupiter, and even "An Account of a Very Odd Monstrous Calf." Responsibility for the publication passed through successive secretaries and remained an unofficial role until 1752, at which time the Society itself assumed editorial and financial responsibilities.

Philosophical Transactions is important not only because it was the first scientific journal, but also because it has set many standards that are still used today. One example is the principle of scientific priority, which states that the person or persons who first perform an experiment, put forward a theory, or make a discovery, should receive the credit for it. Another is that of peer review, whereby the validity of a new scientific work is checked by experts in the field, who also ensure that it is suitable for publication.

Throughout its history *Philosophical Transactions* has published many important papers and letters, such as Isaac Newton's first paper, "New Theory about Light and Colours" and James Clerk Maxwell's "On the Dynamic Theory of the Electromagnetic Field."

ISAAC NEWTON PUBLISHES *PRINCIPIA*

Sir Isaac Newton (1642–1726) is one of the most famous and celebrated physicists of all time. With the publication of *Principia* he revolutionized physics by cementing in place the modern scientific method, championing empiricism, and offering an explanation for much of the world around us through his laws of motion and gravitation.

Isaac Newton spent much of his youth on his home farmstead of Woolsthorpe Manor with his grandmother. At around the age of twelve he was sent to The King's School in Grantham, central England, but his mother later pulled him from education so that he could work the farm. Newton despised farming and eventually, with the help of the school headmaster, was able to return to complete his schooling. He became a top student and was admitted at Cambridge, where he later received a scholarship that allowed him to secure a masters degree. While still studying, he solved some prominent mathematical problems, and in the process developed calculus, the form of mathematics used in almost all physics for the next several hundred years and still widely used today.

In 1665 the university was briefly closed due to the Great Plague, and Newton returned home. There he performed some of his greatest experiments and wrote up his theory of optics, which first brought him to the attention of the scientific community. It was during this time that he supposedly saw an apple fall from a tree, which is said to have inspired his concept of gravity, although the truth of this story is questionable. Upon his return to university he became a fellow of Trinity College, Cambridge, taking up a teaching role, where he taught, among other things, the optics he had written about. He took over as Master of Trinity College and was elected a Fellow of the Royal society in 1672. In around 1679 Newton began serious work on the mechanics of moving bodies, during which time he used his newly developed calculus to perform calculations, although this wasn't used in the final publication. Instead, he reverted to a more widely used form of mathematics. He also studied Kepler's laws of planetary motion and was inspired by the appearance of a comet in 1680, which he wanted to be able to explain.

ICON: A painting of Sir Isaac Newton from 1702. He remains one of the most famous and respected scientists of all time.

A MOMENTOUS PUBLICATION

Philosophiæ Naturalis Principia Mathematica (more commonly referred to as *Principia)* translates to "Mathematical Principles of Natural Philosophy" and is perhaps a suitably grand title for what is considered one of the most important works in the history of science. It was first published on July 5, 1687, through the Royal Society, and almost immediately changed the way physics was done.

Principia is split into three books. The first, *Of the Motion of Bodies*, looks at objects with no resistive forces, such as friction or air resistance. It mostly deals with imaginary and unrealistic scenarios, but the science it presents also describes objects in space, such as planets and moons. It very neatly proves Kepler's laws of planetary motion (see *De Revolutionibus*, pages 36–39) and several other astronomical ideas, such as the fact that spherical objects (including stars or planets) can be mathematically treated as though all of their mass exists at a tiny point at their center, which makes many calculations far easier.

The second book is a continuation of the first. It introduces resistive forces not addressed in the first book and covers the motion of many different things, including pendulums, waves, and even light. This was the most practical of the books, but it is the one most often ignored or forgotten, because many of the major points it makes, such as that light acts as solid particles called corpuscles, have since been disproved or improved to a significant degree. The book was also mostly a direct challenge to many of the ideas presented by René Descartes (1596–1650). Newton showed that following Descartes' ideas to their logical conclusions provided results that simply did not match what was observable in the universe. According to Newton, for a theory to be correct it must match observation exactly.

In the third and final book, *On The System Of The World*, Newton presented his theory of universal gravitation, which states that every particle in the universe attracts every other particle. He went on to describe the consequences of this law, and in so doing

NEWTON'S THREE LAWS OF MOTION

LAW 1 *Every body perseveres in its state of rest, or of uniform motion in a right line, unless it is compelled to change that state by forces impressed thereon.*
An object that is moving will stay moving unless acted upon by an external force.

LAW 2 *The alteration of motion is ever proportional to the motive force impressed: and is made in the direction of the right line in which that force is impressed.*
Force is a product of the mass of an object and its acceleration.

LAW 3 *To every action there is always opposed an equal reaction; or the mutual actions of two bodies upon each other are always equal, and directed to contrary parts.*
For every action, there is an equal and opposite reaction.

solved the irregularities in Earth's orbit around the sun, which are explained by the fact that both bodies orbit around a "common center of gravity" that is slightly off the center of the sun. The book also endorsed the heliocentric model and was, scientifically at least, the final nail in the coffin of the geocentric model. Newton's universal law of gravity gave rise to the idea that some laws might be applicable throughout the universe, and it also created the classical field theory, which led to the formulation of electromagnetism (see Clerk Maxwell's laws, pages 96–99)—a concept that has underpinned many modern scientific advances. The calculus that he developed has become the basis for almost all mathematics used in physics today. You'd be hard pressed to find a scientific paper that doesn't use differentiation, integration, or some other form of Newton's calculus.

In *Principia* Newton characterized the discovery of the underlying universal laws as the ultimate goal of physics. He also finalized the modern scientific method, which Galileo had helped develop, which holds that any discrepancy between theory and observation, no matter how small, renders the theory incorrect, or at least incomplete, and that a better explanation should be sought. *Principia* represents the peak of the scientific revolution, the point at which modern physics emerged as a distinct science. It had established the scientific method, which was fully accepted by most scientific communities, and the real scientific work could begin.

Newton famously wrote in a letter to English polymath Robert Hooke: "If I have

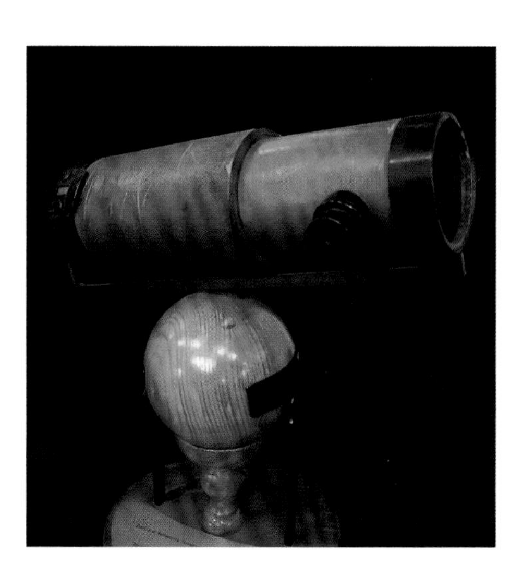

NEWTON'S TELESCOPE: A beautifully preserved replica of the second reflecting telescope that Sir Issac Newton built in 1672.

seen further it is by standing on the sholders [sic] of Giants." This is an incredibly insightful and humble remark by the man. It encapsulates the idea that science is done by many people over many lifetimes, and it continually builds on the work of others.

The importance of *Principia* lies in the fact that it managed to overturn common knowledge, introduce a number of new laws, and set physics apart as an independent science. Thanks to this, the work became the cornerstone of physics. From astronomy and hydrodynamics to mechanics and metaphysics, almost everything in the following 200 years of physics relied heavily in some way on the things Newton laid out in *Principia*.

DANIEL GABRIEL FAHRENHEIT INVENTS THE MERCURY THERMOMETER

Temperature is and has always been a particularly tricky thing to define and measure. Daniel Gabriel Fahrenheit (1686–1736) managed not only to create a device so accurate and easy to use that it is still in many homes today, but also to introduce the first standard temperature scale.

Temperature is, in its simplest form, the measurement of how much energy the atoms in a material have. It is something humans have an inherent sense of, but for a long time measuring it was near impossible. To work, a thermometer relies on two physics principles: expansion in heat and thermal equilibrium.

Any object or substance will increase in size as it is heated and shrink as it is cooled. Normally, this change is so small that it is pretty much unnoticeable. It is possible to see its effects in our daily lives, such as cracks in sidewalks and streets, which can be caused by repeated expanding and contracting through

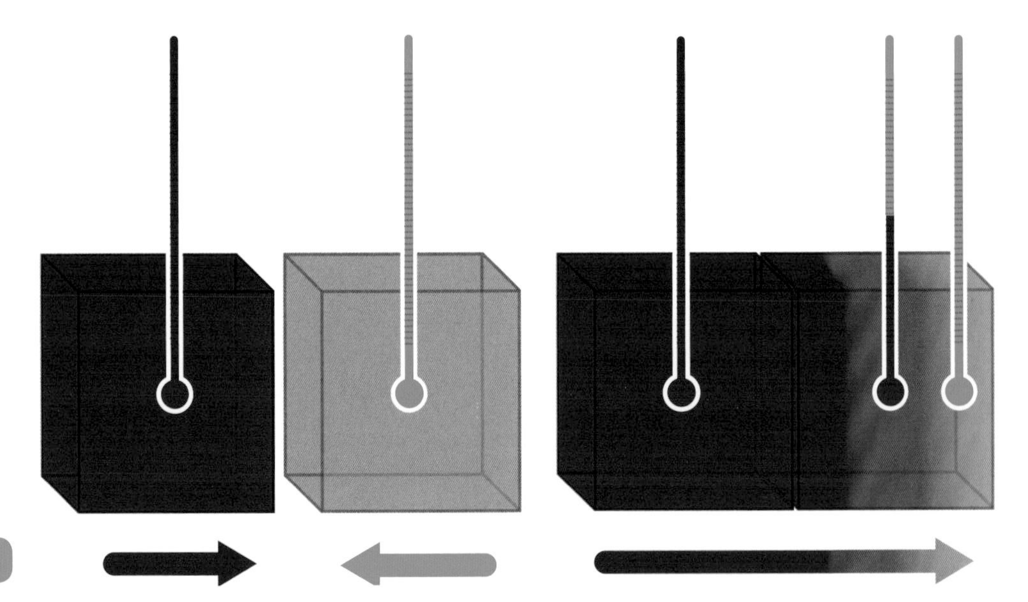

the day–night cycle. The law of thermal equilibrium states that when two objects are in contact, heat will flow from the hotter object to the colder one until they are both at the same temperature. This is easily seen in the world around us; for example, leaving a cup of tea out for too long will cause its heat to enter the air of the room it is in, thus cooling the tea down. Equally, an ice cream will be heated by the warmer air of a room, causing it to melt.

A thermometer exploits these two laws by taking a substance with a known rate of expansion, which means we know how much larger it will get as it is heated up. We then place it in contact with something we wish to measure the temperature of. After a time, the two things reach the same temperature and we measure the increase in size of our testing substance, which it is then possible to convert into a temperature change.

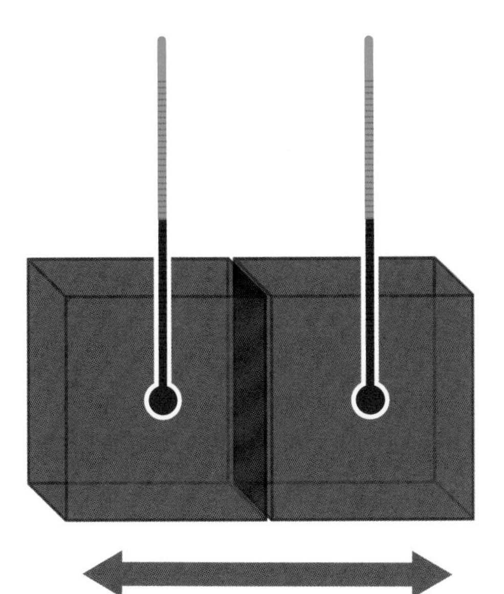

THERMAL EQUILIBRIUM: An illustration showing how thermal equilibrium works. The energy is transferred from the hotter object to the colder one, until they reach the same temperature.

THE FIRST THERMOMETERS

Thermometers were not new in Fahrenheit's time. The first recorded thermometer was created by Galileo in 1593, but it was a simplistic water-based device that could only tell you if an object got hotter or colder, or stayed the same. In 1612 Italian physician Santorio Santorio improved this by creating a device with a scale on it that allowed the specific amount of change to be measured. However, this device used air, which also expands and contracts as a result of changes to atmospheric pressure, thus making it very inaccurate. In 1654 the Grand Duke of Tuscany solved this problem by sealing the measuring material inside a glass tube. This not only protected the device from pressure changes but allowed use of the device in substances such as liquids. However, he used alcohol as his measuring substance, which was also very unreliable.

Daniel Gabriel Fahrenheit initially trained as a merchant in Amsterdam, but he settled in The Hague and worked as a glassblower. Unhappy with the inaccuracy of thermometers, he set about attempting to improve them. He settled on using mercury as the core, whose expansion is easier to predict than alcohol, and he used the latest glassblowing techniques to create a more precise casing. The result was the most accurate thermometer ever created. It was such an effective device that the thermometers used today are much the same as the original design.

SETTING THE STANDARD

While the mercury thermometer was a significant achievement, Fahrenheit's greatest contribution to physics came with the seemingly common sense creation of a standard scale. His eponymous scale was based on three relatively new fixed points of temperature. 0°F is the point at which a mixture of ammonium chloride, water, and ice reached equilibrium. 32°F is the temperature of water just as ice begins to form on its surface. And 96°F is human body temperature, which is usually measured by placing the thermometer into the mouth. While these don't sound particularly scientific, they were easy to achieve in a lab and meant that everyone could now work at the same temperature without difficulty.

This might not seem all that impressive, but having a standardized scale is incredibly important in science. Imagine trying to buy a cabinet to house your television without having a standardized system of

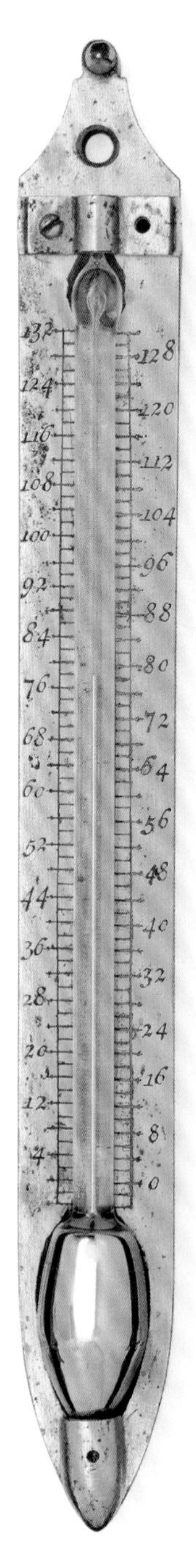

FAHRENHEIT'S THERMOMETER: A Fahrenheit thermometer with a mercury-filled glass core.

measurement, instead using your hands to measure the cabinet. It is likely that the cabinet will be too big or small because the shop owner has different-sized hands to you. By using a standardized measuring scale, such as the metric system of centimeters and meters, this can be avoided.

Although it seems absurd now, hand spans were used as a measurement in ancient Greek and Egyptian times. But if, for example, you start to use a new compound that the instruction tells you will ignite at 50 degrees, and you use a different temperature scale without realizing it, there could be explosive consequences.

The standardization that Fahrenheit introduced was followed by a boom in experiments based on temperature, such as its effects on various materials and their interactions. Temperature can also affect experiments that aren't specifically measuring it, by altering variables such as reaction speed, resistance, and pressure. For this reason, scientific papers began noting the temperature of the area in which the experiment was performed. Modern experiments still use this in STP (standard temperature and pressure) which is set at 0°C and 1 atmosphere of pressure.

THE CHANGING SCALES

In 1724 Fahrenheit reworked his scale, using slightly more precise reference points, such as the boiling point of mercury at 300°F, and making the difference between the freezing and boiling point of water 180 degrees. This made all sorts of equations and calculations easier and was quickly adopted as the new standard. Alternative scales have since been created, of which the most widely used today is the centigrade scale. It puts the freezing point of water at 0°C and the boiling point at 100 °C, which fits significantly better into the now widely used metric system. The scientifically accepted scale, however, is the kelvin. It sets 0°K at absolute zero, which is the lowest temperature it is possible to reach (around -273°C).

The centigrade scale is a modified version of the Celsius scale, which was created by the Swedish Anders Celsius. While the words "Celsius" and "centigrade" are used interchangeably, the two scales are different. Both, however, place the freezing point of water at 0°C and the boiling point at 100°C.

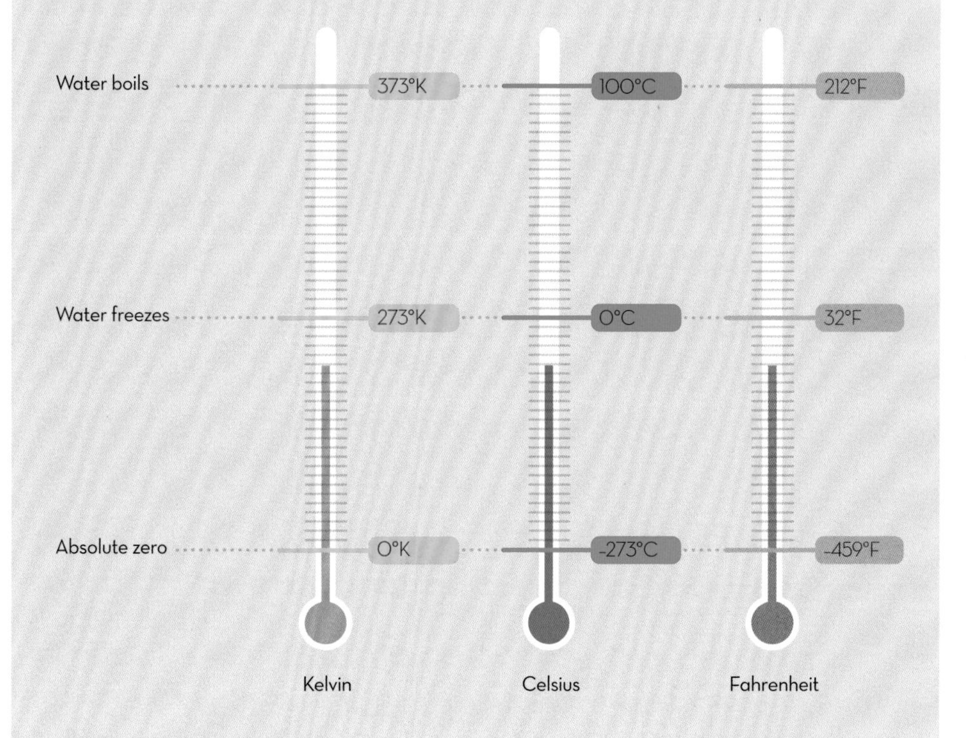

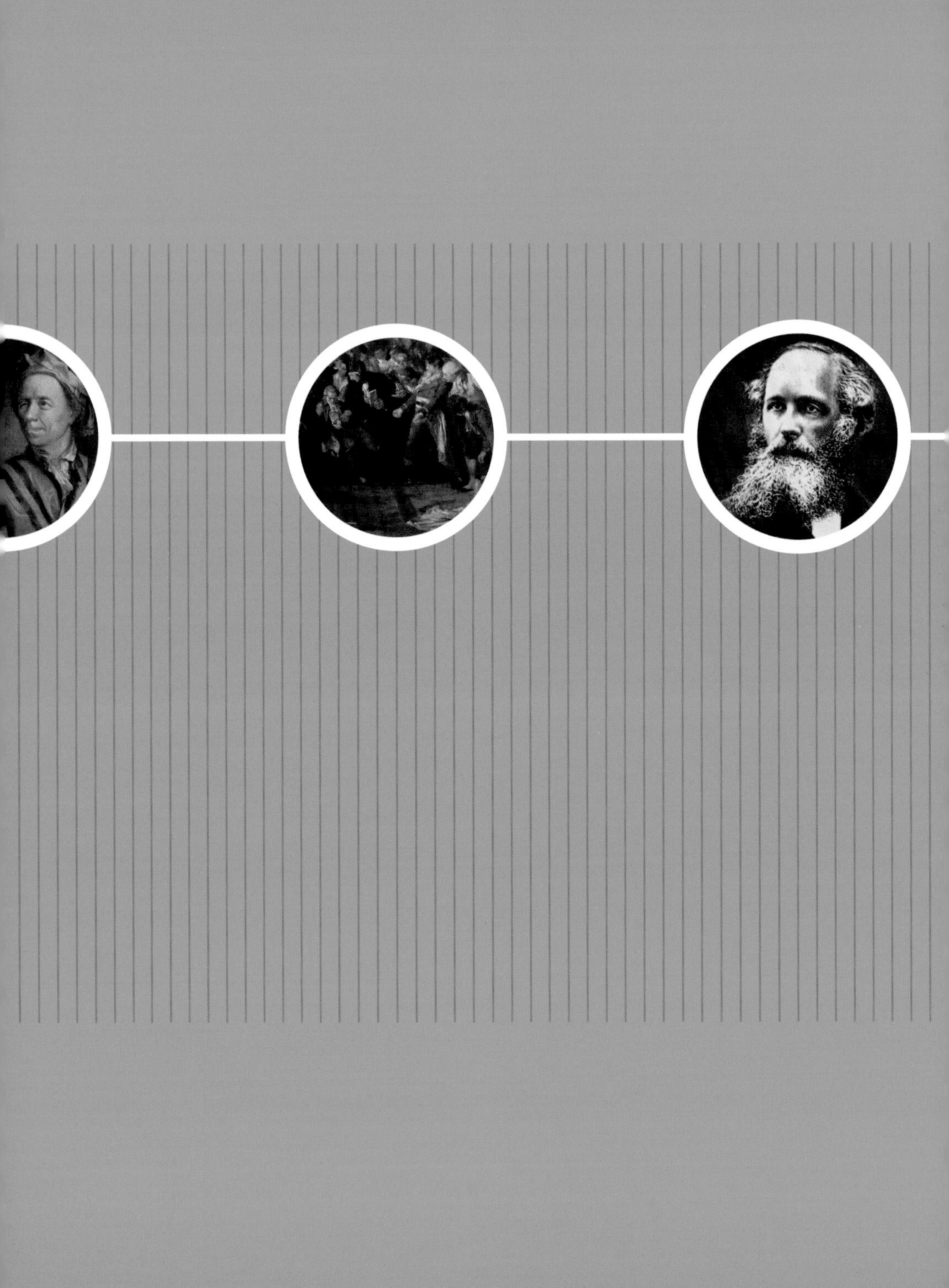

3

CLASSICAL PHYSICS

EULER'S IDENTITY IS PUBLISHED

Physics is inextricably linked to mathematics. Time and again we see the underlying laws of physics throw out the same few numbers, and the importance of this is not lost on us. Swiss mathematician Leonhard Euler (1707-1783) employed some of these numbers in his famous equation, known as Euler's Identity, which is often cited as an exemplar of mathematical beauty. It shows us the inherent interconnectedness of mathematics.

At first glance, Euler's Identity might not seem too impressive, nor its function wholly clear, but don't let that fool you.

$$e^{i\pi} + 1 = 0$$

It combines a number of basic mathematical operations and constants in a wonderfully concise form. However, before we look at what makes this equation work, we need to very quickly define *e* and *i*.

e

e is an irrational number, which means that it doesn't end. It is what you get when you try to calculate the equation below:

$$1 + \frac{1}{1} + \frac{1}{1\times2} + \frac{1}{1\times2\times3} + \frac{1}{1\times2\times3\times4} + \cdots + \frac{1}{1\times2\times\cdots\times\infty} = e$$

This is an infinite series called the Taylor Series, which comes up a lot in physics. Many of Newton's ideas of calculus were based on these infinite sum series and *e* plays a vital role in them. Because it is infinite, it can't be properly calculated, but we are able to find the first trillion or so digits, and most calculations are accurate using the first three or four. NASA, for example, uses only the first sixteen (2.7182818284590452).

e occurs in numerous places, such as when calculating forces and vibrating objects, and even in relativity.

i

i is the solution to −1. It is an imaginary number, meaning it represents a mathematical outcome that is impossible according to normal mathematical systems. So we use *i* to represent it, which allows us to perform calculations that would otherwise not be possible, including those relating to the motion of waves, which is fundamental to quantum mechanics, heat transfer, and optics.

π

π is one of those symbols that you have almost certainly seen before. It is a special number that holds a very important place in mathematics. It is the ratio of a circle's circumference to its radius—that is to say, if you were to measure the point from the center to the edge of a circle and then flatten the circle out into a line, the ratio between the two would be π. This works for any circle; no matter how big or small, it will always be exactly π (3.1415926...) times farther around the outside than from the edge to the middle.

HIGH PRAISE: A mid 18th-century engraving of Leonhard Euler, creator of what has been called "the most remarkable formula in mathematics."

Therefore π features in lots of mathematics, especially when you deal with circles and angles.

WHY IS EULER'S IDENTITY SO SPECIAL?

Given that *e* and *i* represent somewhat abstract mathematical concepts, the importance of Euler's Identity comes from its ability to link them to more recognizable, everyday math that most people are familiar with. The equation allows for the conversion of complex functions (answers that contain an imaginary part) into usable things. Its actual real-world function is to transform Cartesian coordinates (such as are found on a normal *x* and *y* graph) to polar coordinates (a system based on the angle and radius of a circle connecting to a point on a graph), which allows for easier plotting of many types of data. And the equation itself physically rotates an object 180° around the center of the graph.

The five constants in Euler's Identity—*e*, *i*, π, 1, and 0—are also the most common constants in physics. Any calculation or formula you use in any part of physics will likely include one of them, and here we find them all linked together in a single equation. To link such a concise and simply stated equation to so much of the mathematics we use, and with it so many physics concepts, is a significant achievement—one that takes us a step closer to discovering a formula that can describe everything. It's no wonder then that the great Richard Feynman once called it "the most remarkable formula in mathematics."

HALLEY'S COMET RETURNS AS PREDICTED

Halley's Comet is a spectacular sight that appears in our night sky every 76 years or so. It has been observed since at least 240 BC, but it wasn't until 1705, and the work of English astronomer Edmond Halley (1656-1742), that the timing of its return could be accurately predicted.

Comets have been a frequent occurrence throughout history, and they were often seen as omens or signals from the gods. Early astronomers often thought that they were caused by some kind of event occurring in the atmosphere. However, Tycho Brahe's work on parallax (see page 40) showed that they had to be at least farther away from Earth than the moon.

In 1705 Halley produced the paper "Synopsis of the Astronomy of Comets," in which he used Newton's recently published laws (see pages 54–57) to calculate the effect of Saturn and Jupiter on the comet. This meant that he was able to work out an approximate orbital period (the time it takes for the comet to orbit the sun once) of about 76 years. By doing this, he was able to identify three sightings of a comet in the records as being the same comet returning. With this knowledge he was able to predict that the comet would next appear in 1758.

WHAT ARE COMETS?

Comets are made mostly of dust and ice—which has led to their being called "dirty snowballs"—plus a handful of other chemicals, such as methane, ammonia, and carbon dioxide. Their sizes vary; they can be as small as a boulder or as large as a city (Halley's Comet is about ten miles [16km] long, five miles [8km] wide and five miles deep).

Like other bodies in space, comets are pulled into orbit around stars by gravity. However, unlike when approaching objects such as planets and asteroids, a comet's approach toward a star causes it to warm up and this causes the ice to begin melting. As they melt, material is released that forms the distinctive tails comets are known for. While comets are often depicted as moving with their tail behind them, the tails will in fact always point away from the sun, regardless of the direction of the comet's motion.

The comet was first spotted on December 25, 1758, by an amateur German astronomer. The comet became visible to the naked eye in mid-March the following year. This slight alteration in the estimated appearance date of the comet was caused by small perturbations in the orbits of Jupiter and Saturn, which were not known at the time of Halley. They were, however, calculated by other astronomers shortly before the comet's appearance in 1759. Regardless of the relative inaccuracy of his prediction, Halley had managed to successfully predict the comet's return, proving that comets orbit the sun and also providing solid proof of Newton's laws. Unfortunately, Halley died in 1742 at the age of 85, so didn't live to see his prediction come true. But, in recognition of his achievement, the comet now carries his name.

THE HISTORY OF HALLEY'S COMET

Prompted by the knowledge that the comet returns periodically, people have searched through historical records for earlier sightings. The first recorded sighting occurred in 240 BC in China, with appearances also recorded by the Babylonians in 164 BC and 87 BC. Almost every sighting of the comet since has been recorded by somebody somewhere in the world, including famously in the Bayeux Tapestry, which recorded the Norman invasion of England in 1066.

The comet last appeared in the skies in 1986, although at a size almost four times smaller than on earlier occasions. The European, Japanese, and Russian space agencies sent probes to study it, giving us more information about the formation and composition of comets than ever before. This research provided valuable information on the composition of comets and the nature of their surfaces, which the European Space Agency used while planning their *Rosetta* space probe, whose *Philae* lander module touched down on the comet 67P/ Churyumov–Gerasimenko in late 2014. Halley's Comet's next appearance in our skies is forecast for July 28, 2061, where it will have a magnitude of –0.3 (see box, page 71), which would make it brighter than most of the stars in the night sky. Its next reappearance will be in 2134, when it will be almost ten times closer and have a magnitude of about –2.0, which would make it brighter than the planet Jupiter.

HALLEY'S COMET: An image of Halley's Comet taken in 1986 from Easter Island. A comet's tail will always point away from the sun.

JOHN GOODRICKE EXPANDS THE HEAVENS

We all know that the stars above us twinkle, but many astronomers since the ancient Greeks have known that the brightness of some stars physically changes. The English astronomer John Goodricke (1764–1786) made it his mission to discover why this happens and in doing so discovered orbiting stars and expanded our understanding of the size of our universe.

John Goodricke was born into minor nobility in Yorkshire, England. In his early years he fell ill with scarlet fever and was left deaf. He was sent to the Thomas Braidwood Academy, a school for deaf pupils in Scotland, where he developed a love of mathematics and astronomy. After his time there he returned to Yorkshire and lived in the city of York. He quickly developed a friendship with his neighbor, Edward Piggott (1753–1825), through their shared love of astronomy, and they both took an interest in so-called "variable stars," stars with variable brightness. Edward's father owned the largest and most sophisticated private observatory at the time, which Goodricke made extensive use of.

Much of John Goodricke's work focused on Algol A, a star in the constellation of Perseus whose variable brightness had first been noted in 1670 by Geminiano Montanari (1633–1687). After much observation, Goodricke put forward two possible explanations for how this might occur. The first was that the star had a darker side, or some kind of dark area on its surface (perhaps a concentration of sunspots), which produced a periodic dimming of the star as it rotated. The second was that the star was being orbited by a very large but significantly less bright object, and it was this theory that was found to be correct.

BINARY STARS

Stars are made in huge dust clouds known as stellar nebulae, and with so many stars born in the same region of space, it's not surprising that a couple of stars can begin to orbit a common center of mass. While systems of more than two stars can occur, they are often more unstable and collapse quickly, meaning that they are much more rare.

If the orbit of the two stars is on the same plane as the line of sight to Earth, it is possible for one star to move in front of the other and block out the light coming from the object behind. When one star passes in front

A SHORT LIFETIME'S ACHIEVEMENT: A painting of John Goodricke from 1785. Sadly, his life ended prematurely, before he could be informed of his appointment to the Royal Society.

of the other then it will cause the amount of light reaching Earth to drop in the same way a solar eclipse does. The frequency of these drops in light tells us how quickly the two stars are orbiting one another, and the extent of the drop in light can tell us about the sizes of the two stars.

The changing brightness of Algol A, the star Goodricke studied, is caused by this effect. It is actually in a multistar system with two other stars, but the third star, Algol C, is more than 40 times farther away from the two main stars than they are from each other, and it doesn't lie on the same plane, so it doesn't affect the brightness. The primary star, Algol A, is about four times the size of our own sun and 100 times brighter. The second star is more than five times the size of the sun, but only three times brighter.

CEPHEID VARIABLE STARS

John Goodricke looked at a large number of these variable stars and quickly realized that the changes he saw were not all caused in the same way. One of the stars he looked at was Delta Cephei, which sits in the constellation of Cepheus, the magnitude of which varies between 3.5 and 4.4. It soon became clear that the variation wasn't caused by a dark spot or by eclipsing binary stars, and so he gave it the classification "Cepheid variable."

Cepheid variable stars are huge, metal-rich stars that are often supergiants—stars larger than 100 times the diameter of the sun, which can be up to 50,000 times more luminous. What makes them really special, however, is that they pulse. Due to instabilities

COMMON FOCUS: An image of how binary star systems manage to remain in a stable orbit. Notice how both stars orbit around a single central point rather than each other.

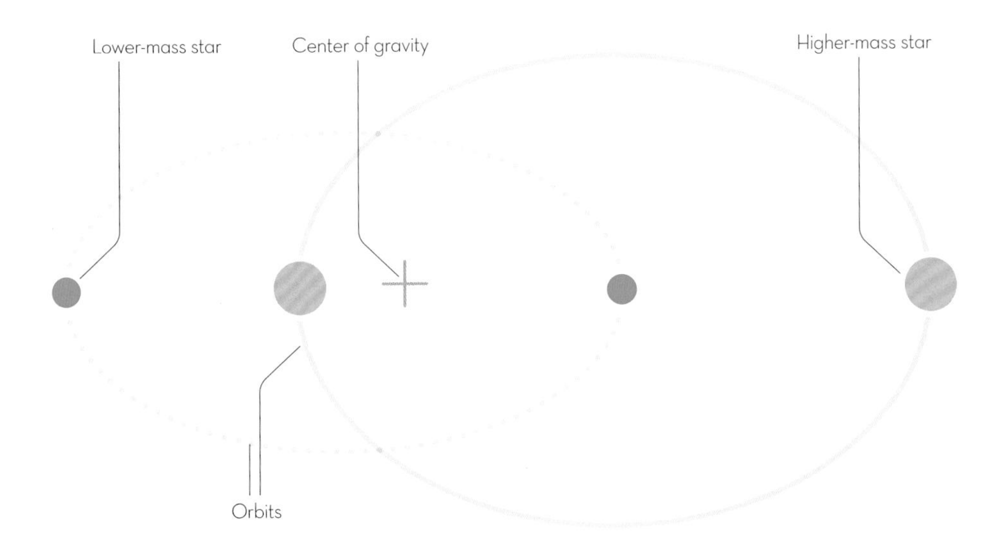

WHAT IS MAGNITUDE?

Magnitude is a measure of how bright an object is. The smaller the value, the brighter the object is. Objects with a magnitude of 5 and above are not bright enough to be seen by the human eye and therefore a telescope is required in order to observe them. Jupiter has a magnitude of -2, and the full moon has a magnitude of -13. On Earth we see Algol's usual magnitude as 2.1, but as Algol B passes in front of Algol A this drops to 3.4 for a period of about 10 hours. This light drop occurs every 2.86 days so we know that the orbital period of Algol B around Algol A is 2.86 days.

in their hydrogen burning, they shrink and grow larger again, and some of them lose up to a quarter of their diameter and over half of their volume with each pulse. This variability in their size and volume means that their magnitude also changes. While this alone is awe-inspiring, what's truly amazing is that there is a direct link between how long it takes for the star to pulse and how bright it is.

Figuring out how far away something is in space can be very tricky, because the distances are so large. Parallax, the method of looking at slight variations in the positions of stars six months apart, on either side of Earth's orbit, is useful, but it only works for objects up to about 300 light–years away. There are also a handful of other methods, but they are often imprecise and only give good estimates. When we look at a Cepheid variable star, however, we can measure precisely how long it takes to vary in magnitude, from which we can calculate how bright the star actually is. The distance of the star from Earth can then be determined by looking at how bright it appears, compared to how bright it actually is.

From these calculations it was realized that the universe is much larger than had originally been thought: Eta Aquilae, a star discovered by Piggot, was found to be 1,400 light–years away, and S Vulpeculae a massive 11,000 light–years away. Today we use methods that allow us to measure even greater distances, such as Type Ia supernovas, which always give out the same amount of light when they explode and can be seen for millions of light–years, or red shift caused by a galaxy's motion away from us but it was Goodricke's work that begun it all.

GOODRICKE'S TRAGEDY

For his work on binary stars John Goodricke was appointed a Fellow of the Royal Society on April 16, 1786, but the news never reached him. Four days after he was elected, but before he had been informed, he died of pneumonia, which it is thought developed as a result of him spending long hours in the observatory at night. He was only 21.

THE METRIC SYSTEM IS INTRODUCED IN FRANCE

Standardized systems of measurements may not fire the imagination, but they represent a crucial development in the history of physics. It can be near impossible to use another scientist's work, if you must first convert all their data into the type you use—a problem that the introduction of the metric system sought to resolve.

The metric system comprises decimal units, which are based on the number 10, so a kilogram consists of 1000 grams, and there are 100 centiliters in a liter. The system was first proposed by the British scientist John Wilkins (1614–1672) as a "universal language of science."

In 1795, the French revolutionary government passed an act that set in law a new measurement system. The new design was based on the decimal system, deemed to be more democratic, so it could be used by "all people for all time," and not just the elites in society.

DECIMAL TIME (TIME) The day was split into 10 hours, consisting of 100 minutes, which had 100 seconds. A week was 10 days long, and there were 12 months, consisting of 3 weeks and 5/6 additional days each year to keep the total of 365 days.

GRADE (ANGLE) Each quarter of a circle was split into 100 grades (also known as gon), making a full circle 400 grades.

METER (LENGTH) One meter is equal to one ten-millionth of the distance from the North Pole to the equator through Paris.

GRAM (WEIGHT) The mass of one cubic centimeter of water.

FRANC (CURRENCY) 100 centimes and 10 décimes to the franc.

Conversion to this new system was poorly planned and hastily implemented, and it was widely unpopular. It was halted in 1812, and the country returned to the old imperial model, but was reintroduced in 1837, after it had become widely used by scientists.

THE SPREAD OF THE METRIC SYSTEM

France had fully adopted the system by 1795, and, despite a few troubles over the next 100 years, it slowly spread throughout much of Europe and the French Empire. It then took root in the Middle East, Russia, and China, and by the time SI units were introduced in 1960, almost the entire world had adopted the metric system.

REVOLUTION: A scene from the French Revolution, which was a major factor behind the adoption of the metric system.

SI UNITS

You have no doubt noticed that the units mentioned before don't match up with what we now think of as the metric system. This is because the metric system was adapted over the years to include more types of measurements, such as energy and temperature, and some of the existing units were changed to make them easier to use for the rest of the world. The modern metric system is the Systéme International d'Unités, or SI units. While it kept much of the decimalization for units like meters and grams, it also adopted other units, such as seconds and angles. It also defines units by far more accurate measurements, such as a meter being the distance light covers in $\frac{1}{299,792,458}$th of a second. The development of this system and its subsequent acceptance across the world has facilitated the spread of scientific progress internationally.

HENRY CAVENDISH CALCULATES G

In his 1687 work *Principia*, Isaac Newton defined gravity as the unknown constant G—the gravitational constant. More than 100 years later, British scientist Henry Cavendish (1713–1810) calculated it.

Newton's law of gravitation was written mathematically as follows:

$$g = \frac{Gm_1m_2}{r^2}$$

Where g is the force of gravity between two objects, m_1 and m_2 are the masses of the objects, and r is the distance between them, which is squared. This leaves G, the gravitational constant, which, as the name implies, never changes. Its use is largely mathematical, allowing for the conversion of the masses and distance to the gravitational force in the correct units. Henry Cavendish was the first to actually perform the experiment necessary to calculate it.

After studying at the University of Cambridge, Cavendish moved to London. His father was a man of great importance within the Royal Society, enabling Henry Cavendish to attend meetings, lectures, and dinner parties held by the society. The younger Cavendish was elected as a fellow in 1760. He became an active member and was appointed to a number of posts, including the Royal Society Council. His scientific work was varied and he became famous in his day for the discovery of hydrogen, which he called "inflammable air."

The Cavendish experiment was designed to allow him to calculate the density of Earth and therefore its mass. Using the experiment, he calculated the density to be 5.448 grams per cubic centimeter, whereas the same calculation run today, using far more precise equipment, gives a result of 5.51 grams per

DID CAVENDISH ACTUALLY CALCULATE *G*?

Cavendish didn't set out to find *G*, nor did he explicitly state it in his paper, so did he actually calculate it? Some have argued that he didn't. Yet because a value for *G* was reached through his calculations and experiment, many prominent physicists, including Richard Feynman (see page 150–153), have credited Henry Cavendish with its discovery.

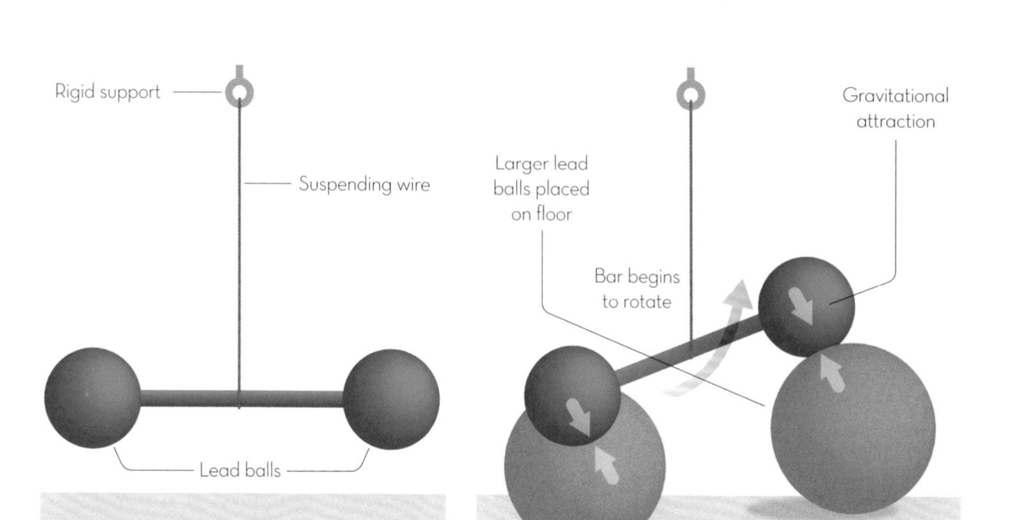

cubic centimeter. When converted into Earth's mass and plugged into Newton's equation, we get G as 6.74×10^{-11}, against today's value of 6.73×10^{-11}.

THE CAVENDISH EXPERIMENT: A diagram showing the equipment that was used for Cavendish's experiment. This setup was enclosed in a wooden shed to protect it from external forces.

THE EXPERIMENT ITSELF

The experiment consisted of a bar suspended from a very thin string, with two lead balls attached to either end. Much larger lead balls were placed on the ground near the smaller balls. The gravitational attraction of the larger balls on the smaller ones caused the bar to rotate very slowly until the amount of torsion (the force required to twist something) in the string matched the force of gravity. The torsion was calculated according to the angle that the bar rotated to, and this equated to the force between the two sets of balls. When combined with the weight of the smaller ball, the ratio of the two forces was used to calculate the density of the Earth as around 5.45 times that of water, which was a known quantity, and thus the density of the Earth could be roughly calculated.

The amount of force that twisted the string was incredibly tiny—less than 150,000,000th the weight of the smaller ball—so the experiment had to be very carefully controlled to make sure that the wind or anything else didn't interfere. To do this, the experiment was built inside a wooden box inside a shed that was only open through two small holes, where Cavendish placed telescopes to be able to observe what was going on inside. It was probably the most accurate experiment of its time.

YOUNG'S DOUBLE-SLIT EXPERIMENT IS PERFORMED

The nature of light was the focus for one of science's greatest debates—was it a wave or a particle? Sir Issac Newton championed the idea that it was a particle. However, there were many who were unconvinced and thought that it was a wave. In 1802 Thomas Young (1773–1829) created an experiment which would prove that light does take the form of a wave.

Thomas Young was by all accounts a remarkable man. Born in Somerset, the eldest of a large family, by his early teens he was able to speak fourteen languages. He studied medicine in London and Edinburgh before completing a degree at Cambridge. In 1799 he set up his own practice in London's West End. Throughout much of this time he also worked on many of his own scientific papers and ideas. In 1801 Young was appointed as a professor of natural philosophy at the Royal Institution, where he delivered lectures over two years, which were published in 1807 in *Course of Lectures on Natural Philosophy and the Mechanical Arts*.

IS LIGHT A WAVE OR A PARTICLE?

In 1800, Young presented a paper to the Royal Society that identified light to be a wave. The paper was received skeptically, yet Young was undeterred. He continued working on his ideas, using water to demonstrate his wave principles. In 1803 he performed what has become known as the double-slit experiment, using light instead of water, and his results proved that light is a wave.

Waves can interact with one another through a process called interference. Think of waves on an ocean's surface: they consist of peaks and troughs—points where the water is higher and points where it is lower. Imagine two sets of waves traveling across the ocean that then collide with each other: where one peak meets with another peak, the overall height of the water will increase to accommodate all the water. Equally, where two troughs meet each other, the dip will become twice as low. Constructive interference is the name given to two waves merging to become larger. Conversely, destructive interference describes the meeting of wave and trough, whereby the height of the resulting wave is less than that of the initial wave.

When we throw a stone into a lake or pond we see the waves spread out from it in a circular pattern. If we were to drop two in

THOMAS YOUNG: Despite challenging one of the most respected scientists of the era, his mathematical rigor ensured he was proven right.

at the same time to create our ripples, we would notice the waves interfering with one another.

Now, if we were to stand at the shoreline and measure the height of the water that reaches us, we would notice a pattern caused by the interference. The point equidistant from the two stones will always be the highest peak and then either side would be a low trough. The next peak would be slightly lower and the next trough slightly higher, with this pattern continuing until the effect of the ripples is too small to notice. The pattern is visible regardless of the distance between the two stones or how far from the shore it is. We call the resulting wavefront (the pattern of peaks and troughs at a single point in time) of the shoreline a diffraction gradient.

POLYMATH

Thomas Young wasn't only a great physicist, but also a respected doctor who developed safe practices for child dosages and applied much of his wave theory to explain how blood works. He also studied the eye, describing astigmatism and trichromatic vision.

Remarkably, he also made a significant contribution to translating the Rosetta Stone (an ancient Egyptian text inscribed on stone, found in 1799), his work likely providing the platform for Jean-François Champollion (1790-1832) to fully decipher it in 1822.

With the knowledge that these wavefronts occur for all types of waves, Young set up a single light source, which he directed at a metal sheet with two slits in it. The light would pass through the slits and hit a wall beyond the metal sheet. In view of the fact that the same amount of light would pass through each hole (acting as the dropped stones), if light were indeed a wave then the effects of constructive and destructive interference should be visible on the wall. It was expected that this would manifest as a pattern of light and dark spots.

GRADUAL RECOGNITION

Young's experiment did produce the expected result, thus confirming that light is a wave. Yet despite this, he was publicly attacked for it, and his reputation as a serious scientist suffered. Consequently, the publishing of his lectures was severely delayed and he resigned from the professorship, turning his attention back to his medical practice. The evidence was clear, however, and the theory slowly gained acceptance. One of the crowning moments during his theory's rise in popularity came during the 1817 French Academy of Science awards. In what is possibly an apocryphal tale, French scientist Augustin-Jean Fresnel (1788–1827) delivered an address on his wave theory that combined Young's double-slit experiment with some earlier work of the Dutchman Christiaan Huygens (1629–1695). However, Siméon Denis Poisson (1781–1840), a defender of the particle theory of light, did some quick calculations and after the talk stood up and declared that if what Fresnel said was true

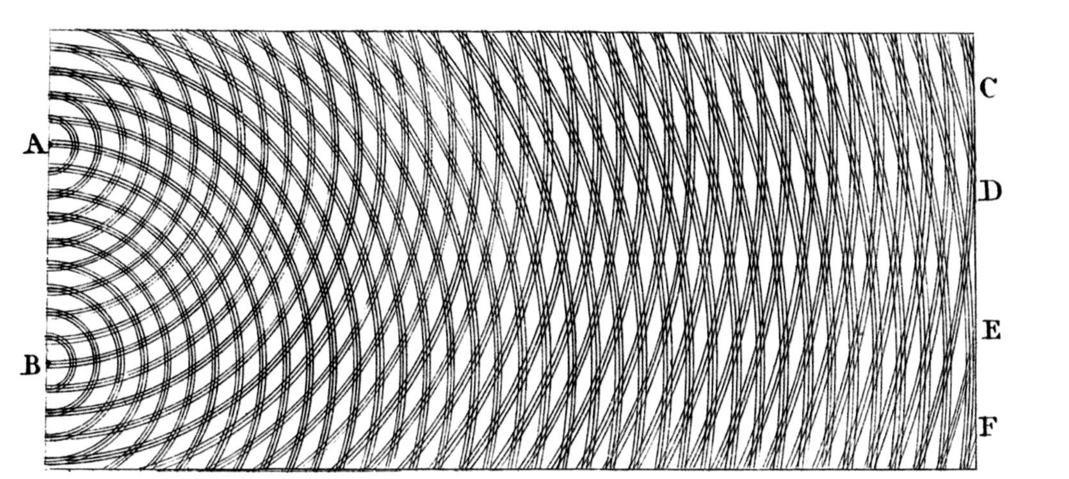

DOUBLE-SLIT EXPERIMENT: The effects of the Young double-slit experiment—a pattern of dark and light spots along the wall, shown here by the letters C, D, E, and F. A and B denote the light sources.

then upon blocking light with a circular disk the shadow of the disk should have a bright spot appear in its center. He called the idea preposterous and, satisfied he had destroyed the idea, sat back down. However, as soon as he had, another academic stood up and revealed that they had observed this very effect and were due to present it later in the proceedings. Poisson was proved wrong by the appearance of the bright spot, and it was named the "Poisson spot," adding insult to injury. After this incident, the wave theory of light became the accepted model.

A WAVE AND A PARTICLE

The wave theory of light was excellent and explained much about the properties of light and how and why it acted in certain ways, but there were a few things that it just couldn't solve. Most notable of these was a phenomenon known as the photoelectric effect (see page 114), whereby an electrical circuit could be made by using light of a high-enough wavelength to expel electrons from a sheet of metal. If light were a wave, then any wavelength of light could be used (as long as the intensity was increased accordingly), but this was not the case. Albert Einstein solved this in 1905, by proving that light was a particle. So now there was proof that light had properties of both waves and particles! The resolution to this apparent contradiction came with the discovery of wave-packets called photons, which are able to act as waves or particles, depending on the circumstances.

JOHN DALTON DEVELOPS ATOMIC THEORY

Understanding how chemicals and elements work is an interesting and important question. While atoms were a commonly accepted idea in the late 1700s, their properties were basically unknown. It was John Dalton (1766-1844) who took the first big step in understanding them.

John Dalton was born in 1766 near Cockermouth, England. He showed academic promise, but his status as a Quaker, and therefore not part of the Church of England, saw him prevented from attending most universities. He was eventually offered a teaching position in a special academy for religious dissenters in Manchester. His early work was largely focused on meteorology, but he also studied color-blindness in detail (from which he suffered), as a result of which the condition was for some time known as Daltonism.

Dalton's ongoing work on the composition of gases led to his discovery of the law of multiple proportionality, which states that when combining two elements into a compound they will always combine in the same ratio. This means that if 10g of

JOHN DALTON. Despite the fact that his atomic theory suffered at first from a lack of evidence, it later became the leading theory for how elements and molecules are made.

carbon bonds with 3g of oxygen then 20g of carbon will bond with 6g of oxygen, and so on. By 1804 Dalton had done so much experimentation with his gases that he created a theory for why it happened. This atomic theory had four main principles:

- Elements are made of atoms.
- Atoms of any given element are identical to each other.
- Atoms of elements combine in compounds in whole number ratios.
- Atoms can't be made or destroyed; any chemical reaction is some form of combination, separation, or rearrangement of atoms.

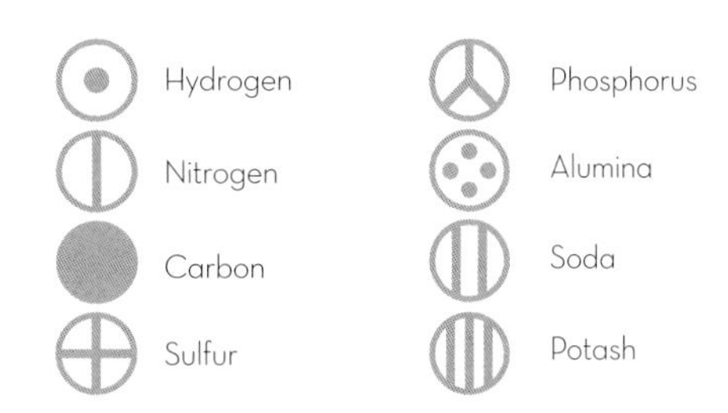

DALTON'S SYMBOLS: A number of the symbols proposed by Dalton in his *New System of Chemical Philosophy*. These were subsequently replaced by a system of letters, which is still used today.

THE ATOMIC WEIGHT

Dalton also began to weigh chemicals relative to one another. He had been using hydrogen, oxygen, carbon, and nitrogen in different mixtures and was able to calculate their approximate weight. He found that compounds of the correct proportionality would weigh the same, which led him to his idea of whole number compounds. For example, an amount of carbon (weight 5) would weigh the same as an amount of magnesium (weight 20), provided that the former measure is four times larger than the latter.

He produced a table showing the weights of various elements and compounds, which he first fully published in his 1808 textbook *A New System of Chemical Philosophy*. The table orders elements according to their atomic weight, so that hydrogen, with an atomic weight of 1, is first. This is followed by carbon (spelled "carbone" at the time) with an atomic weight of 5, oxygen with 7, sulfur with 13, iron with 38, and so on. By combining these elements, the relative atomic weight of compounds could be calculated. A water molecule, for example, (which was thought to comprise one oxygen and one hydrogen atom), would have an atomic weight of 8; alcohol, containing three carbon atoms and one hydrogen, would weigh 16.

Today we know that Dalton's work was fundamentally flawed, due in large part to a contemporary misconception about atoms and errors in the understanding of chemical composition (as we have seen with water). His work was improved greatly by the Italian scientist Amedeo Avogadro (1776–1856) in 1811, when he calculated atomic masses far more accurately and realized that many gases were diatomic (they naturally occur as two combined atoms). Atomic theory was further improved by Ernest Rutherford and Niels Bohr (see pages 116 and 120), who explained the atom itself. However, despite its shortcomings, John Dalton's theory was essentially correct, and it became the basis of most early atomic physics.

CARNOT DESCRIBES THE PERFECT ENGINE

Heat and temperature are among the most important, and confusing, fields of study in physics. By the early 1800s steam engines were at the forefront of industrial innovation, as scientists worked to unlock the full potential of the new technology. Sadi Carnot (1796–1832), the "father of thermodynamics," provided the key.

Frenchman Nicolas Léonard Sadi Carnot was the son of a high-ranking member of Napoleon's government and war cabinet. He was a good student and successfully passed the entrance exam to the prestigious École Polytechnique at the age of sixteen—the earliest possible age of entry. Carnot went on to join the engineering corps, before transferring to the General Staff Corps, where he remained on call for military duty. It was during this time that he began attending courses and lectures across Paris, taking an interest in a great many scientific disciplines, most notably industrial machinery and the workings of gases. There was one object that combined the two and took his interest more than any other: the steam engine.

A STEAMY ISSUE

Steam engines were widely used across the industrialized world. Their ability to turn heated water into mechanical motion was used for many things, from spinning cloth to forging iron. Despite the technology's prevalence, not much research had been conducted in France on the steam engine and the French were rapidly falling behind England. The leading theory for how heat worked was the incorrect caloric theory, which described heat as a fluid that repelled itself, which is what caused it to spread out.

The problem with steam engines in Sadi Carnot's time concerned their huge inefficiency. Efficiency is the measure of the amount of available energy produced by a process compared to the amount of it that is actually usable. For example, a light bulb takes in electricity and produces light but it also produces wasted heat and even a small amount of sound. Even with nearly 100 years of innovation, the best still only operated at 5% efficiency.

Carnot looked at two major questions: Is there a limit to how efficient an engine can be? And can an engine run on something other than steam? He published his answers in *Réflexions Sur la Puissance Motrice du Feu* (*Reflection on the Motive Power of Fire*; 1824), which, aimed at the general populous,

SADI CARNOT: Leaving behind a difficult career in the military, Carnot channeled his love of science into trying to solve one of the great problems of the time.

contained very little mathematics, addressing the more complex matters in the footnotes.

THE CARNOT HEAT ENGINE

In his book, Carnot describes an idealized engine—an imaginary machine that is free from the kind of energy loss that would occur in a real engine. It consists of two plates (metal strips) and one piston, and it operates according to the Carnot cycle.

The Carnot cycle and engine operate with the greatest efficiency possible. This represents a perfect heat transfer. By doing some calculus it is possible to establish an

THE CARNOT HEAT CYCLE

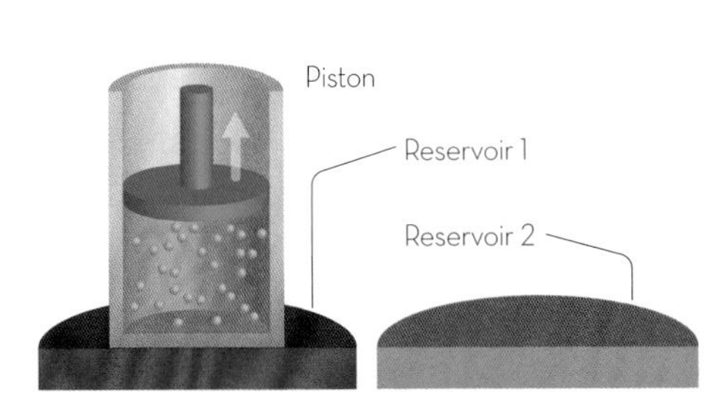

STEP 1 The piston is placed in contact with a thermal plate at temperature T_1. The piston is raised slowly, causing the gas to expand and it is heated at the same rate, such that there is no temperature change. This process causes heat energy to flow from the reservoir into the piston. This is called the isothermal expansion stroke, because the gas is expanded and the process is isothermal (ie, there is no temperature change).

STEP 2 The piston is removed from the thermal plate and continues to expand the gas. No heat leaves or enters the piston, so the gas cools to T_2. This is called the adiabatic expansion stroke, because the gas is still expanding and the process is adiabatic (ie, there is no heat transfer).

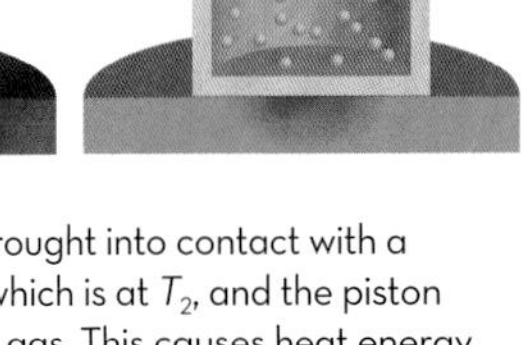

STEP 3 The piston is brought into contact with a second thermal plate, which is at T_2, and the piston begins to compress the gas. This causes heat energy to flow out of the piston and into the thermal reservoir. This is called the isothermal compression stroke.

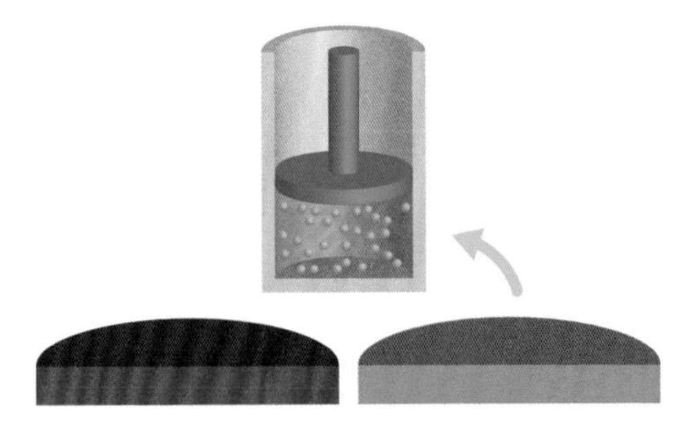

STEP 4 The system is reset by removing the piston from the second plate and compressing the gas, thus causing its temperature to rise to T_1, at which point the process begins again (see Step 1). This is called the adiabatic compression stroke.

THERMODYNAMICS

Thermodynamics emerged from Carnot's efforts to further understand heat. It was developed by a number of prominent scientists, including William Rankine (1820–1872) and Lord Kelvin (1890–1895). The laws of thermodynamics are as follows:

THE ZEROTH LAW If two systems are in thermal equilibrium with a third system, they are in thermal equilibrium with each other.
THE FIRST LAW Energy is always conserved within a closed system.
THE SECOND LAW Entropy in a closed system always increases or remains the same over time.
THE THIRD LAW The increase of entropy will drop to zero if a system is at 0°K.

The first question that you are probably asking is, "What is a 'Zeroth Law?'" It was a law developed after the first three had already been established, and subsequently identified as more fundamental than the others. But rather than sensibly moving all the laws down a slot, the new law was added as the 0th law.

The second question is probably, "What is entropy?" This is a much more difficult question to answer. In simplistic terms, it is the amount of disorder in a system. Imagine you had a box half filled with colored balls that were all grouped together according to their color. If you then shook the box (the equivalent of adding heat), you would see that the balls have become disorganized and are no longer grouped. This is a basic explanation, but defining it formally involves a lot of heavy calculus and mathematics.

equation for the efficiency of the engine as:

$$e_c = 1 - \frac{T_2}{T_1}$$

Where e_c is the efficiency of the Carnot engine and T_1 and T_2 are the temperatures of the two thermal plates (whereby T_1 is always hotter than T_2). This equation tells us that in order to increase the efficiency of the engine we want to increase the difference between the temperatures we use. In order to have 100% efficiency, we would need to reduce the temperature of T_2 to 0°K, absolute zero, which is impossible. We can also see that if T_1 and T_2 are the same temperature, the system will not work at all because the efficiency will drop to 0%. It also tells us that, regardless of the material we use, the maximum efficiency cannot be improved.

Most real engines operate with a T_2 of room temperature (about 300°K) and T_1 at 500°K, giving the maximum possible efficiency to be 0.4 or 40%. Their actual output is much lower than this due to friction and other imperfections, such as heat loss through the materials.

MICHAEL FARADAY CREATES THE FARADAY DISK

Our modern world runs on electricity—phones, computers, cars, and countless other everyday items depend on it. Almost all of the electricity produced is created by an electrical generator, and the first of these was built in 1831 by Michael Faraday (1791-1867).

Faraday was born in 1791 in London and received only a basic schooling, though at the age of fourteen he took an apprenticeship at the local bookbinders. During this time he educated himself and began to attend lectures of the Royal Society, especially those of the chemist Humphry Davy (1778-1829). He sent Davy a 300-page book (which he bound himself) of detailed notes on Davy's lectures and the two struck up an academic friendship. In 1813 Davy injured himself in an accident and subsequently took Faraday on as an assistant.

Faraday began his scientific career in chemistry, assisting Humphrey Davy while also conducting independent research, whereby he studied and categorized a great number of chemicals, created an early version of the Bunsen burner, discovered the existence of nanoparticles (tiny metallic particles in a substance), and later discovered the laws of electrolysis (the process of passing electricity through substances).

Faraday's first recorded experiment with electricity was in 1812, when he used a stack of alternating pennies and zinc disks, along with paper soaked in salt water, to generate a continuous current (like an early battery), with which he performed a number of chemical interactions. However, it wasn't

AUTODIDACT: A painting of Michael Faraday from 1842. He taught himself most of what he knew, before becoming an assistant to Humphrey Davy.

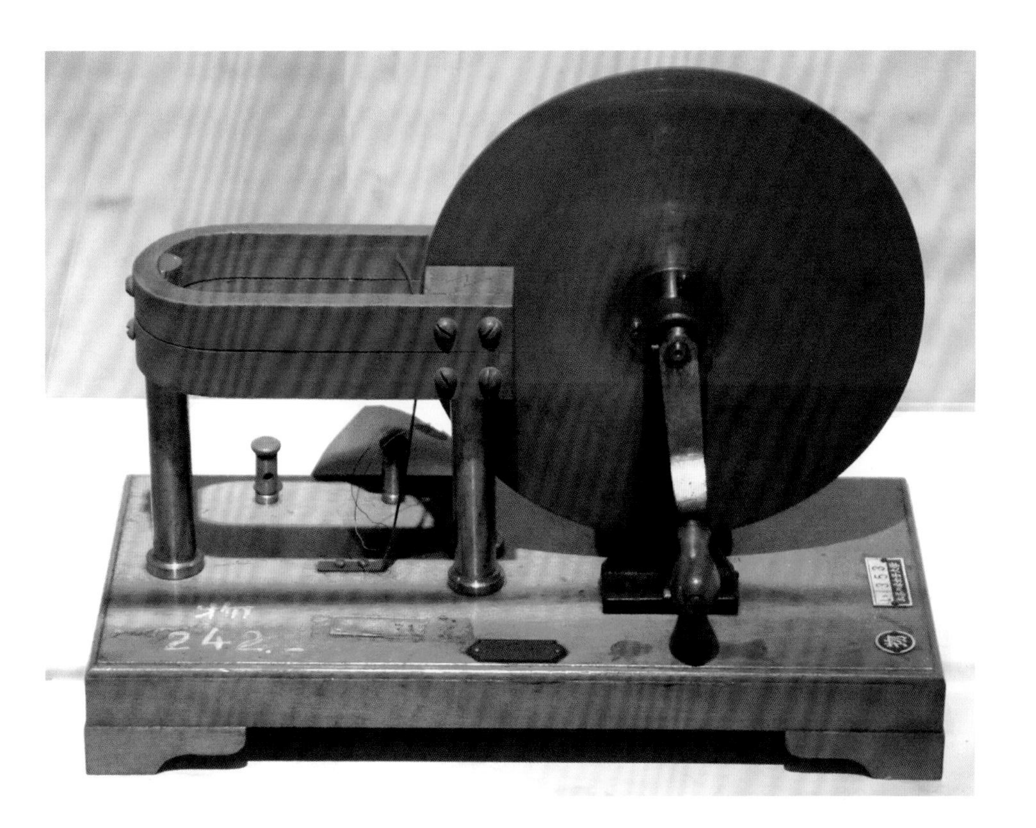

ELECTRICAL GENERATOR: An early example of a Faraday disk. Turning the handle spins the disk and creates a current in the wires.

until 1820, when Hans Ørsted (1777–1851) noticed that the needle of a nearby compass moved whenever a device generating an electrical current was switched off and on, that a link between electricity and magnetism was discovered. In the year following this discovery, Davy, along with another scientist, William Wollaston (1766–1828), began attempting to construct a device that would be able to create motion using the effect. Their attempts to make an electrical motor were unsuccessful, but they discussed their effort with Faraday at great length. Faraday spent much of the next ten years working on optics and researching electromagnetic properties of many materials. However, in 1831, a couple of years after Davy's death, he began working in earnest to create an electromagnetic generator.

THE FARADAY DISK

What Faraday created was the Faraday disk, or homopolar generator. The device consisted

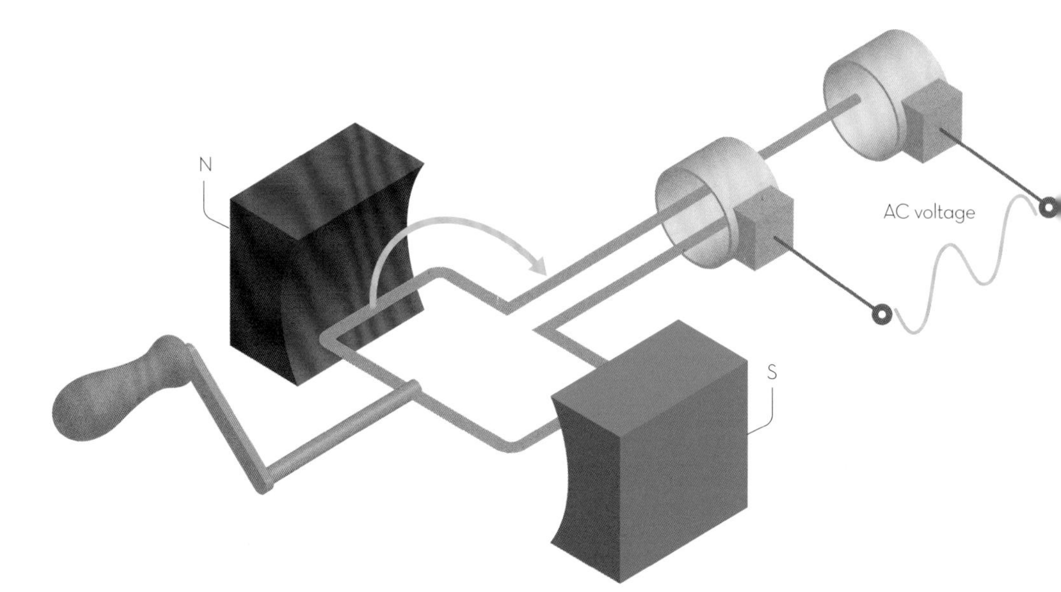

GENERATOR: A simple generator, based on Faraday's initial design. The copper shaft is rotated between magnetic poles by means of a crank.

of a large copper wheel that was attached to a wire on one side and a handle on the other. Part of the disk was then placed inside a strong magnet and attached to the other end of the wire by metal brushes, which allowed the wheel to turn freely while remaining in contact. Through the action of a handle, the wheel is spun, causing parts of it to pass through the field and generating a current in the wires. To explain this phenomenon, he developed the Faraday law of induction:

$$\nabla \times E = -\frac{\partial B}{\partial t}$$

What this equation states is that a changing magnetic field will create an electric field, and vice versa. It shows that the electrical current within the wire is produced when the disk passes through the field, with the changing force inducing the current. Also, because it is possible to reverse the process by applying a current to a system, it is possible to make the wires move. This is how electric motors, such as those in electrical cars and electrical toothbrushes, work.

What Faraday had done was produce the first mechanical means of generating electricity. His invention essentially made electricity both safe and viable for use outside of the science lab. Modern versions of Faraday's initial design are responsible for producing the power used everywhere.

IMPROVING THE GENERATOR

The Faraday disk is actually an incredibly inefficient way of generating electricity. The main reason for this is that the large copper disk, while producing the useful current, also makes many "eddy currents," which are caused by tiny imperfections in the material and flow in the opposite direction to the main current, thus reducing its overall effect massively.

Hippolyte Pixii (1808–1835) made the first major improvement to the generator, replacing the wheel with a spinning shaft that offered double the current and using an alternating current (AC) instead of direct current (DC), which reduced the eddy currents hugely. This meant that more of the current was able to flow through the wire, thus increasing the amount of energy that could be generated.

One of the first true induction motors was patented in 1887 by Nikola Tesla (1856–1942). This generated AC. AC is where the current constantly switches between moving backward and forward, whereas DC is where the current flows in only one direction, producing a single constant voltage. AC became the preferred form of electricity, as it is easier to produce. Additionally, with the use of transformers, the voltage can be changed. This is especially useful when transporting it across large distances through power lines, where having the current traveling at high voltages reduces the amount of power lost. Electrical lines can carry current at voltages in excess of 30,000 volts, which would be incredibly unsafe to have in homes. A transformer can be used to reduce the voltage to safe levels for everyday use.

The largest improvement came when it was realized that the current was generated when the field was "cut" by the wire (you can test this yourself by passing a wire hooked up to a voltmeter between the poles of a horseshoe magnet) and that the more times the wire passed through the field the more electricity would be generated. This led to the wire being wrapped around the spinning shaft. Each coil would cause more to be generated; two coils would produce twice as much electricity as a single coil, and ten coils would make ten times as much, and so on. Simple generators, such as hand-powered torches that utilize a crank to produce the rotations, tend to have around 1,000 coils of wire, and generators in power plants contain millions.

MODERN POWER STATIONS

Almost all modern power stations use the method established by Faraday to produce electricity. All fossil fuel power stations (such as coal or gas) burn their fuel, which is then used to heat up tanks of water to boil. The steam then rises up the tank and on its way up pushes a turbine. This causes the turbine to rotate. The turbine is then connected to a dynamo, which rotates within a magnetic field to generate huge amounts of electricity. Some renewable power sources, such as wind turbines, use the same method, harnessing natural elements to produce the rotations.

HAMILTONIAN MECHANICS IS CREATED

We use mathematics to describe physics, and sometimes the type of mathematics we use to look at a given problem can make the problem easier or harder. There have been many systems for mathematics, but none of them come close in importance to the introduction of Hamiltonian mechanics.

Physics is written in the language of mathematics. Sometimes it can be hard to express an idea or feeling in our own language, so we turn to another to do so. *Déjà vu*, *faux pas*, and *schadenfreude* have all been incorporated into the English language to represent ideas for which there had been no English words. So it is in physics that using different types of mathematics (the equivalent of different languages), things can be made much easier.

Newtonian mathematics is built on the basis of forces. If you were to use it to calculate an interaction between two objects, you would look at how the forces work with or against each other. This is the perfect way of looking at how large-scale systems such as planets work, and it is even a good fit for many simple interactions. However, it doesn't take much before the equations become very complicated. For example, the mathematics of a mass swinging on a pendulum, which has another pendulum hanging off, involves a great many second-order differential equations, which needless to say can be time-consuming and difficult to work with.

Lagrangian mechanics, created in 1788 by Joseph-Louis Lagrange (1736–1813), is based on energies rather than forces. For the double pendulum example, Lagrangian mechanics requires you to look only at the gravitational and kinetic energy of each of the masses, as opposed to looking at the tension, gravitational, and resistive forces in all three dimensions (height, width, and depth). It should be noted that the two equations describe exactly the same thing; they are just different ways of doing it.

Hamiltonian mechanics was formulated in 1833 by William Hamilton (1805–1865). Without getting too far into the mathematics, Hamiltonian mechanics allows for two main things, statistical values and generalization of quantized (multiple single-value) states. Why are these important? Because quantum mechanics is an inherently statistical form of mathematics and the results are always quantized. In essence, this mathematical system made quantum mechanics possible. Using the Newtonian mathematical system, it just wouldn't have been possible to work it out. Hamilton's work made it all possible.

MATHEMATICAL PIONEER: A painting of Joseph-Louis Lagrange from ca. 1800. His creation of a new system of mathematics opened the door to lots of new discoveries.

LAGRANGIAN MECHANICS

Lagrangian mechanics involves looking at the energies of a system rather than its forces. But, you may be asking, why is this so important? The reason is that when you get down to the subatomic level almost everything is calculated in terms of energy. Mass, position, speed, and other variables are all directly dependent on the amount of energy present in a system. This means that when our mathematics is based on energies, it becomes much easier to do calculations on any given system.

Hamiltonian mechanics also gives us a new type of answer to a problem. Whereas systems such as Newtonian mechanics will always give you a certain value (such as the number 4 or 7), Hamiltonian mechanics is able to output a set of solutions from one problem. This means it might give you an answer of $3x+1$; as x can be any value, we get solutions that go 4, 7, 10, 13, and so on. It is this ability to produce a set of answers that makes it so important and allows it to be the basis for quantum mechanics.

BOLTZMANN PUBLISHES HIS EQUATION

With the form of mathematics now available thanks to William Hamilton, it was possible for Ludwig Boltzmann (1844–1906) to begin work on statistical mechanics, which deals with properties of atoms and matter, the first step toward quantum mechanics.

Ludwig Boltzmann was born in Vienna and received an early education from a private tutor before going to school in Linz. He went on to study physics at the University of Vienna in 1863 and graduated three years later with his PhD thesis on the kinetic theory of gases. He worked as a lecturer and assistant to Josef Stefan (1835–1893). In 1869 he took up a post as a professor of mathematical physics at the University of Graz.

Through the continuation of his own study of the kinetic theory of gases, he developed and then published the Boltzmann equation (sometimes known as the Boltzmann transport equation) in 1872. It describes statistically the behavior of a thermodynamic system (a system of temperatures) that is not in perfect equilibrium (ie, it is not all the same temperature). The generalized equation can be written as:

$$\left(\frac{\partial f}{\partial t}\right) = \left(\frac{\partial f}{\partial t}\right)_{\text{force}} + \left(\frac{\partial f}{\partial t}\right)_{\text{diffusion}} + \left(\frac{\partial f}{\partial t}\right)_{\text{collision}}$$

LUDWIG BOLTZMANN: His work provided a means of dealing with huge sets of data, which would otherwise have been unmanageable for scientists.

This equation gets very complicated very quickly, for example:

$$\left(\frac{\partial f}{\partial t}\right) = \iint gI(g,\Omega)[f(p_A'^{,t})f(p_B'^{,t}) - f(p_A'^{,t})\, f(p_B'^{,t})]\, d\Omega\, d^3p_A\, d^3p_B'.$$

What this equation does is comprehensively describe all of the possible things that a given particle may do. Its importance lies in the fact that it is not derived by analyzing the position of each individual particle at every moment in time, as in Newtonian mechanics. Instead, it looks at the probability that a number of particles will occupy a small region of space and have an equally small change in momentum in the small moment of time.

This fundamental shift of perspective has, it is fair to say, had a profound impact. It makes possible previously impossible calculations, across fields such as the transportation of heat around objects and the conductivity of materials. This use of the new mechanics to calculate things statistically rather than analytically, as had been done before, meant that many more types of calculations could be done and equations solved. This opened the door to statistical mechanics and eventually to quantum mechanics.

STEFAN–BOLTZMANN LAW

Boltzmann and his teacher Josef Stefan managed independently to deduce a law about the radiative properties of a body. Stefan did it in 1879, and Boltzmann in 1884 through thermodynamics. The equation looks like this:

$$j^* = \sigma T^4$$

It states that the power radiating from a "black body" (a theoretical object that absorbs all incoming radiation) is directly proportional to the fourth power of its temperature. σ is the Stefan–Boltzmann constant that puts the answer into usable units. Its value is 5.67×10^{-8}.

This is an important equation because it allows for the temperature of objects to be measured just from their light output. The first notable application of this was in the measurement of the temperature of the sun. Stefan was able to calculate the sun's temperature to be 5700°K, which compares well to the modern value of 5778°K. This equation could then be expanded to calculate the temperature of other stars in our universe, and it could even be used for moons, planets, and other celestial objects.

MAXWELL–BOLTZMANN DISTRIBUTION

Boltzmann later worked with the scientist James Clerk Maxwell (see pages 96–99) on

THE LAW OF DISORDER

Boltzmann also did much work on the second law of thermodynamics: The total entropy in a closed system will always increase over time. Boltzmann used his equation to develop the H-theorem, which describes a tendency in an ideal gas for the value of entropy (H) to increase. His work provided an excellently practical demonstration of the ideas of entropy, which had been around since the time of Carnot and it also provided a convenient way to visualize entropy, which is an otherwise abstract concept.

It was because of this statistical analysis that the second law of thermodynamics became known as a law of disorder. Imagine a deck of cards fresh out of the box and in a nice neat order. If you randomly shuffle the cards a little, they will begin to move out of order. While it is perfectly possible for you to randomly shuffle your pack of cards so that all of the aces are together, then all of the 2s, the 3s, and so on, the chance of this is astronomically small. It's $1/8.09 \times 10^{67}$ or 1 in 80,658,175,170,943,878,571,660,636,856,403,766,975,289,505,440,883,277,824,000,000,000,000. Therefore in practice this doesn't occur, and the deck, when shuffled, will tend toward becoming disordered. Fascinatingly, due to the enormous number of possibilities, whenever you shuffle a deck of cards they are, probabilistically speaking, in an order that has never occurred before in history.

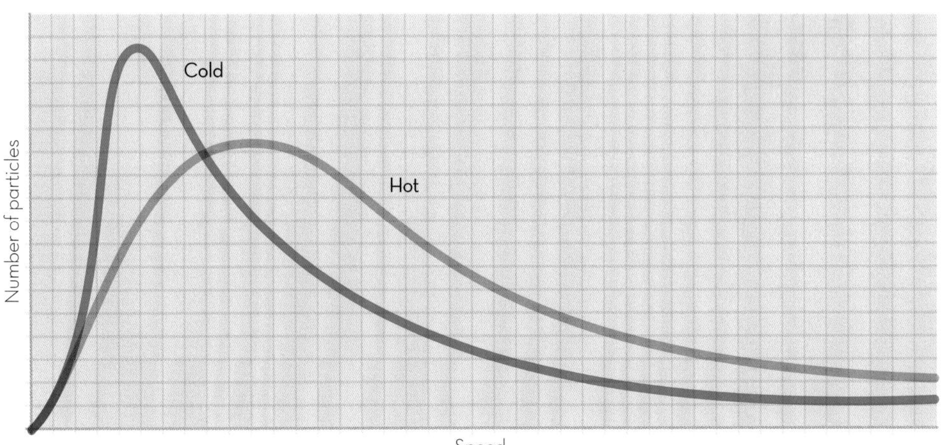

MAXWELL–BOLTZMANN DISTRIBUTION: The Maxwell-Boltzmann distribution, shown here in terms of two different temperatures. The higher temperature produces a broader distribution.

creating a probability distribution of particles in a system. This condenses a lot of numbers into an easily readable graph. The Maxwell–Boltzmann distribution is:

$$f(v) = \sqrt{\left(\frac{m}{2\pi kT}\right)^3}\, 4\pi v^2 e^{-\frac{mv^2}{2kT}}$$

When plotted out for two different temperatures, we get the above graph.

We can see that for a given system of particles there is a peak speed, where most of the particles are found, with a sharp rise up from the lowest speed to that peak. After the peak there is a decreasing number of particles with progressively higher speeds. We also notice that by adding more energy into the system (by increasing the temperature), it is possible for us to increase the speed of the peak of particles, though we do get fewer particles with the peak amount of energy.

It is possible to perform a number of calculations on this graph. For example, you can calculate the most probable speed of any given particle with:

$$v_p = \sqrt{\frac{2kT}{m}}$$

And equally, the expected value of any given particle can be calculated with:

$$\langle v \rangle = \frac{2}{\sqrt{\pi}}\, v_p$$

In this statistical analysis we can begin to see the start of quantum mechanics forming. In our system we are able to extract from it a probability. By using the graph we can state the probability that a particle picked at random will have a given speed, but we cannot say for definite the speed of any one particular particle.

JAMES CLERK MAXWELL REVEALS HIS EQUATIONS

The Maxwell equations are among the most important in physics. They contain everything there is to know about the electrical and magnetic forces, which we now know as a single electromagnetic force. While James Clerk Maxwell (1831–1879) may not have discovered each of them, it was he who presented them collectively in 1873 and demonstrated their connectedness to the world.

James Clerk Maxwell was born in 1831 in Edinburgh, Scotland. At the age of fourteen he wrote his first paper, "Oval Curves," which showed how to draw curves using string, as well as examining the properties of a number of ellipses and ovals. It was deemed worthy of being presented on his behalf to the Royal Society of Edinburgh.

At the age of sixteen he began attending the University of Edinburgh, turning down Cambridge because of his love for his teachers. During this time he devoted himself to much private study, writing a number of papers, including "Equilibrium of Elastic Solids" and "Rolling Curves." He then studied at Cambridge University, graduating in 1854 and immediately thereafter presenting his mathematical work to the Cambridge Philosophical Society. He also applied for a fellowship at Trinity College, where he had studied. He gained his fellowship on the October 10, 1855, and began preparing to teach. However, when his application to the vacant Chair of Natural Philosophy at Marischal College in Aberdeen was successful, he left Cambridge. He remained at Marischal until 1860, working on various problems, including the rings of Saturn, at which time he was rather unceremoniously sacked, when the college was merged with another to form the University of Aberdeen. He then moved to London and became the Chair of Natural Philosophy at King's College London.

It was during this time that Maxwell put serious work into electromagnetism. Using Ampère's circuital law, he was able to calculate the speed of electromagnetism to be equal to that of light, though he failed to recognize the underlying significance—that light is an electromagnetic wave. He began gathering all the knowledge he could on electric and magnetic fields, reducing all the knowledge on the topic to just 20 equations in his work "On Physical Lines of Force," which was published in 1861. He continued his work on electromagnetism, finally publishing

JAMES CLERK MAXWELL: While he may not have discovered any of the equations, his work in unifying them into one force was a huge step.

the four fundamental equations that explain electromagnetism in his 1873 *A Treatise on Electricity and Magnetism*. The equations were originally written as partial differentials, which are not easy to read, and English scientist Oliver Heaviside (1850–1925) converted them to the more commonly used and easily understood differential equations.

THE EQUATIONS

Gauss's law:

$$\nabla \cdot E = \frac{\rho}{\epsilon_0}$$

Gauss's law of magnetism:

$$\nabla \cdot B = 0$$

Faraday's law of induction:

$$\nabla \times E = -\frac{\partial B}{\partial T}$$

Ampère's circuital law:

$$\nabla \times B = \mu_0 J + \frac{1}{c^2}\frac{\partial E}{\partial T}$$

GAUSS'S LAW

$\nabla \cdot E$ represents the amount of electrical field passing out of an object. The equation tells us that the amount of field passing out of an object is proportional to the amount of charge contained within it (ρ) divided by the constant the permittivity of free space.

What this means is that a material's own electric field is only subject to the charge contained within it and is not affected by any outside factors. The constant ϵ_0 represents how well electrical field lines can be produced in empty space.

GAUSS'S LAW OF MAGNETISM

$\nabla \cdot B$ represents the amount of magnetic field passing out of an object. As this value is 0, it means that any magnetic field lines that leave an object must come back into the object. This is how we get the magnetic field shapes on objects like bar magnets. This consequently means that there is no such thing as a magnetic monopole, which in turn means all magnets must have a north and south pole.

FARADAY'S LAW OF INDUCTION

$$\nabla \times E = -\frac{\partial B}{\partial T}$$

This law states that a magnetic field that changes in time $\frac{\partial B}{\partial T}$ is able to produce an electric field $\nabla \times E$. This is induction and is what Faraday used to create his Faraday disk (see pages 86–89).

AMPÈRE'S CIRCUITAL LAW

$$\nabla \times B = \mu_0 J + \frac{1}{c^2}\frac{\partial E}{\partial T}$$

This law states that a wire with a current running through it will produce a magnetic field that is proportional to the amount of current passing through it. This equation also sets the speed of electromagnetic waves at c (the speed of light, 3×10^8 meters per second). The constant μ_0 is the permeability of free space; it is a measure of how well fields are able to interact in empty space.

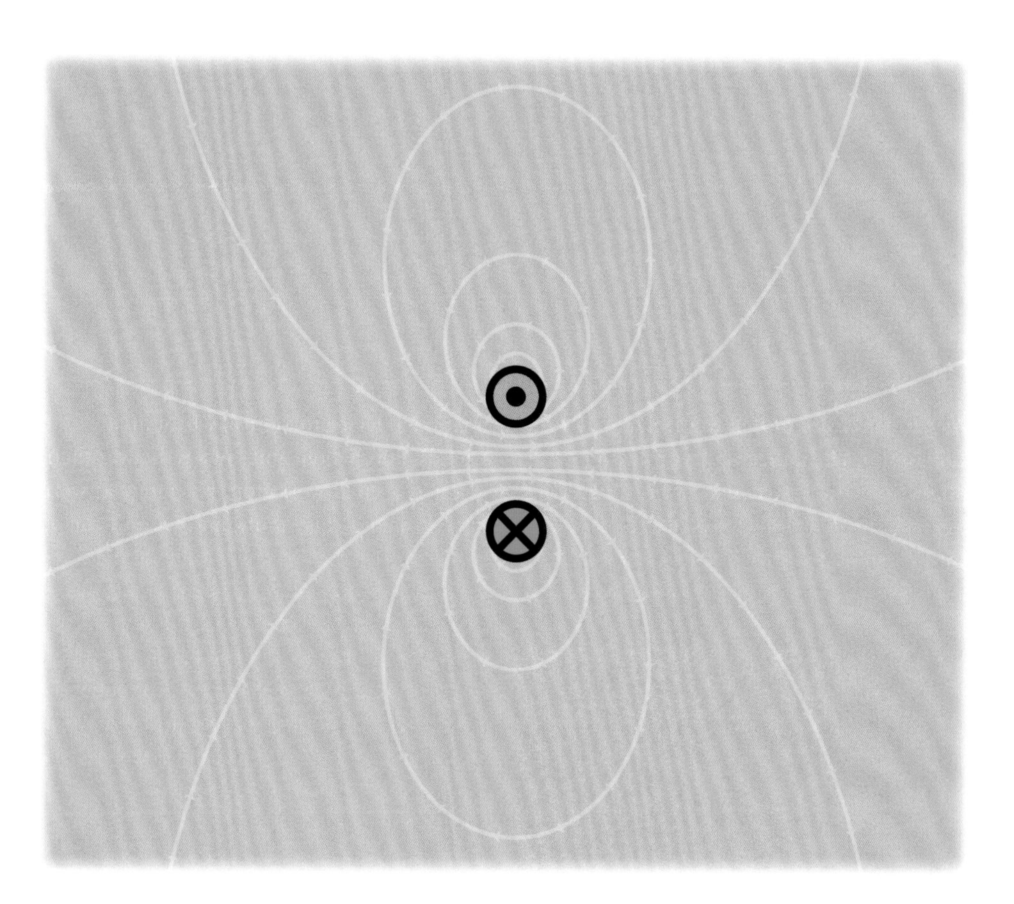

ELECTROMAGNETIC FIELD: An electromagnetic field, which is generated by a moving electric charge, comprises interacting electric and magnetic fields.

THE IMPORTANCE OF THE EQUATIONS

The Maxwell equations are incredibly simple and yet they describe so much of one of the fundamental forces of nature. From them, it is possible to derive almost everything there is to know about the electromagnetic force, from electrostatics to tensor calculus and even some optics.

It was perhaps one of the first pieces of "fundamental" physics work, attempting to encapsulate all our knowledge of a topic within one set of information—something that has become one of the cornerstones of modern physics. Maxwell's work simplified and contributed to almost every electrical development after him. The magnetic principles revealed by the equations are the core of all the electrical devices that power our everyday life.

ALEXANDER GRAHAM BELL INVENTS THE TELEPHONE

The story of the telephone, and of the race to patent it, is perhaps one that you have heard before. Was Alexander Graham Bell (1847-1922) its inventor, or did he steal the design? While this is an interesting tale, of more scientific importance is the phone's function itself.

Alexander Graham Bell was born in 1847 in Edinburgh. His father was a professor of phonetics with a particular interest in trying to teach the deaf how to speak. Alexander showed an early talent for music and invention. By the age of twelve he had created a device to dehusk wheat at a friend's family flour mill, where he was then granted a small workshop. His interests were encouraged by his family, and with the help of his brother he created a "talking head" by using windpipes and bellows, while also experimenting with the family dog to make it appear to talk. By the age of nineteen his interest in sound had taken an academic turn, when he began working on resonance. Upon discovering that his intended research had already been done, he became somewhat disillusioned and focused instead on helping his father. In 1870 the family moved to Canada and Bell (who was often of frail health) found himself reinvigorated and began to experiment with sound and electricity.

In 1857 the Frenchman Édouard-Léon Scott de Martinville (1817–1879) invented the phonautograph, which showed that sound was made of vibrational waves. It created physical representations of the

ALEXANDER GRAHAM BELL: The technology he created quickly became one of the dominant forms of communication across the world.

waves on paper (by using the sounds to cause an arm with a pen to move up and down, in a similar way to a seismograph), which in theory could be converted back to sound. There was much work done with sound before Bell turned his focus to the problem of transmitting voices over a distance.

Bell drew inspiration from the "tin can telephone," by which vibrations are transmitted between two cans along a piece string. The telegraph, a contemporary means of long-distance communication that had been commercially available since the late 1830s, also informed his work. However, the telegraph functioned according to a single on/off signal, which meant that it could only send information in Morse code, thus placing a practical limit on the amount of information that could be conveyed in a single communication.

Drawing on the existing technologies, Bell created a device that employed a flexible reed in an electromagnetic field to almost perfectly replicate sounds. He then adapted this so that the reed was sensitive enough to be moved by a voice, thus producing a replica, albeit a rather poor one, at the other end of the line.

THE SIGNIFICANCE OF BELL'S WORK

In addition to revolutionizing the way people communicated with each other, Bell's invention was the first device capable of the sophisticated transfer of complex information. While today everything is done digitally (through the use of what is basically much faster telegraph), Bell's work not only

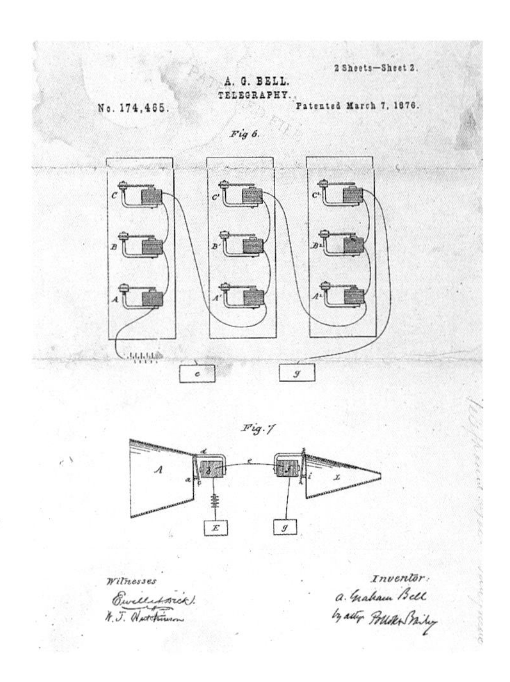

THE ORIGINAL PATENT: Alexander Graham Bell's patent, filed on March 7, 1876. Bell's status as the inventor of the telephone is a matter of dispute.

paved the way for the telephone's creation but also laid the foundation for radio, television, and the next century and a half of analogue communication.

MICHELSON AND MORLEY FIND NOTHING

The truth of science is that, more often than we would like, an experiment proves nothing. This is called a null result. While this might sound disappointing, it can sometimes be just as important as a positive result. One null result in particular stands out as probably the most important of all time.

Edward Morley (1838–1923) was a sickly child and was home schooled until the age of nineteen. He attended Williams College in Massachusetts, where he showed an aptitude for making instrumentation, creating a chronograph to accurately determine the latitude of the college. He was appointed a professor of chemistry at the college in 1868, where he remained until his retirement in 1906.

Albert Michelson (1852–1931) was born in Prussia. At the age of two, his family relocated to Nevada, and Michelson later moved to San Francisco to study, during which time he lived with an aunt. He received a special appointment to the U.S. Naval Academy, where he became interested in science and performed one of the first experiments to accurately measure the speed of light. He later left the navy, taking up a post as professor of physics in Cleveland, Ohio, in 1883. It was upon taking this post that he struck up a working relationship with Morley and asked for his assistance with an experiment he had been unable to complete, one that looked at something called luminiferous ether.

"ETHER WIND": In the late 19th century, it was posited that Earth moved through the luminiferous ether, causing an "ether wind."

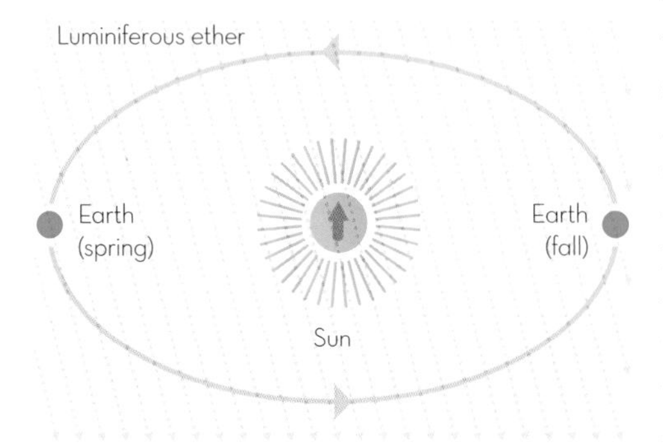

WHAT IS LUMINIFEROUS ETHER?

The now obsolete idea of luminiferous ether isn't easy to understand as it has no real equivalent in modern science. It was thought to be a substance that filled all of space, which provided the medium through which light waves move (in the same way that sound waves move through air). And as light is able to pass through a vacuum, such as space, it was assumed that luminiferous ether must also be present in a vacuum.

AWARD WINNERS: Albert Michelson (left) was the first American to receive the Nobel Prize in Physics; Edward Morley (right) was awarded the Davy Medal for outstanding work in the field of chemistry.

But what was the importance of the ether? Imagine you have three people, Albert (A), Benoît (B), and Copernicus (C). A is sat on a chair in a field, B is on a bus, which is on a road next to the field, and C is on a train whose tracks run next to the road. The speedometer on the bus reads 20km/h (13mph) and the speedometer on the train reads 50km/h (31mph). So when the two vehicles pass by the field, A, who is stationary, sees B traveling at 20km/h and C traveling at 50km/h. From B's point of view, he sees C traveling forward away from him at only 30km/h (as his speed of 20km/h is subtracted from C's speed of 50km/h) and he also sees A moving backward away from him at 20km/h. To add to this confusion, C sees A moving backward away at 50km/h and B moving backward away at 30km/h. Given all of these conflicting viewpoints, who is actually traveling at what speed?

This is complicated further by the idea of different frames of reference, or by light, which always has a set speed (don't worry we'll cover all of this later in the book, on pages 124–127). But at the time of Michelson and Morley, the answer was easy: the ether, which was considered to be an all pervasive substance that filled everything, could be treated as stationary, and everything else could be calculated in relation to it.

THE MICHELSON–MORLEY LAB:
A photograph of the equipment used in the Michelson-Morley experiment.

The significance of ether was precisely that it permitted a constant means of calculating how fast something travels. The speeds of two moving objects relative to each other were of no concern; only the speed of each object next to this stationary ether was of consequence. This made things much simpler. Take the example from earlier: To calculate the speeds of A, B, and C, relative to the speeds of everything else, we must also factor in that Earth moves through the universe at 108,000km/h (67,000mph), that our solar system moves at 828,000km/h (515,000mph), and that the whole galaxy is moving through space. If you do the calculation, you will find that A moves at around 3,000,000km/h, and B and C slightly faster, depending on the speed of their vehicles.

THE EXPERIMENT

With ether central to contemporary physics, Albert Michelson decided to try to measure it. However, after his first experiment was deemed too imprecise to detect the expected changes, he paired up with Morley to try once more.

The experiment consisted of a beam of light fired at a half-silvered mirror, through which half of the light would pass and the other half would be reflected (thus creating two beams). The two light beams would then reflect off mirrors and either reflect off or pass back through the half-silvered mirror into a detector. The two beams of light would then interfere (as shown previously with Young's double-slit experiment; see pages 76–79). It was expected that the light beam that was working against the ether wind (an effect caused by Earth's movement through the ether) would mean that one of the beams would slow down fractionally and the detector would receive a fringe pattern (similar to that produced by Young's experiment) proving the existence of ether.

When they performed the experiment they saw nothing: no fringe or any evidence of the ether. The experiment was mounted upon a mercury bath to allow it to be rotated so that it would find the ether wind, but it still didn't work. The experiment was repeated multiple times throughout the year to try to account for potential changes throughout the year, but time after time nothing was detected.

Eventually, it became obvious that, although there were some small changes detected, they were clearly caused by imperfections within the equipment. So in 1887 Michelson wrote in his paper published in the *American Journal of Science*:

> *The experiments on the relative motion of Earth and ether have been completed and the result decidedly negative. The expected deviation of the interference fringes from the zero should have been 0.40 of a fringe - the maximum displacement was 0.02 and the average much less than 0.01 - and then not in the right place. As displacement is proportional to squares of the relative velocities, it follows that if the ether does slip past, the relative velocity is less than one sixth of Earth's velocity.*

In this rather technical statement is the admission that they had found nothing. They had set out to figure out the speed differences throughout the year, but the experiment had failed to find the ether at all. Further experiments by others in 1903 and 1904 confirmed the lack of ether. This discovery, or rather the lack of one, was significant because it changed how physicists had to think about the universe, leading directly to the development of relativity.

MORLEY–MICHELSON EXPERIMENT: The half-silvered mirror (center) splits the light beam, sending the two resulting beams onward to two mirrors (shown in yellow), and then to the detector.

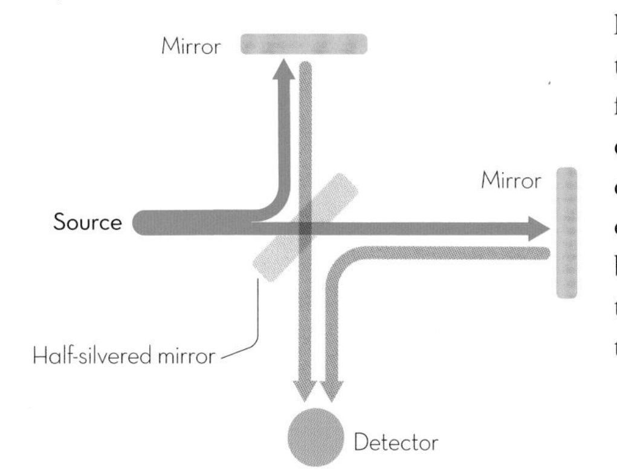

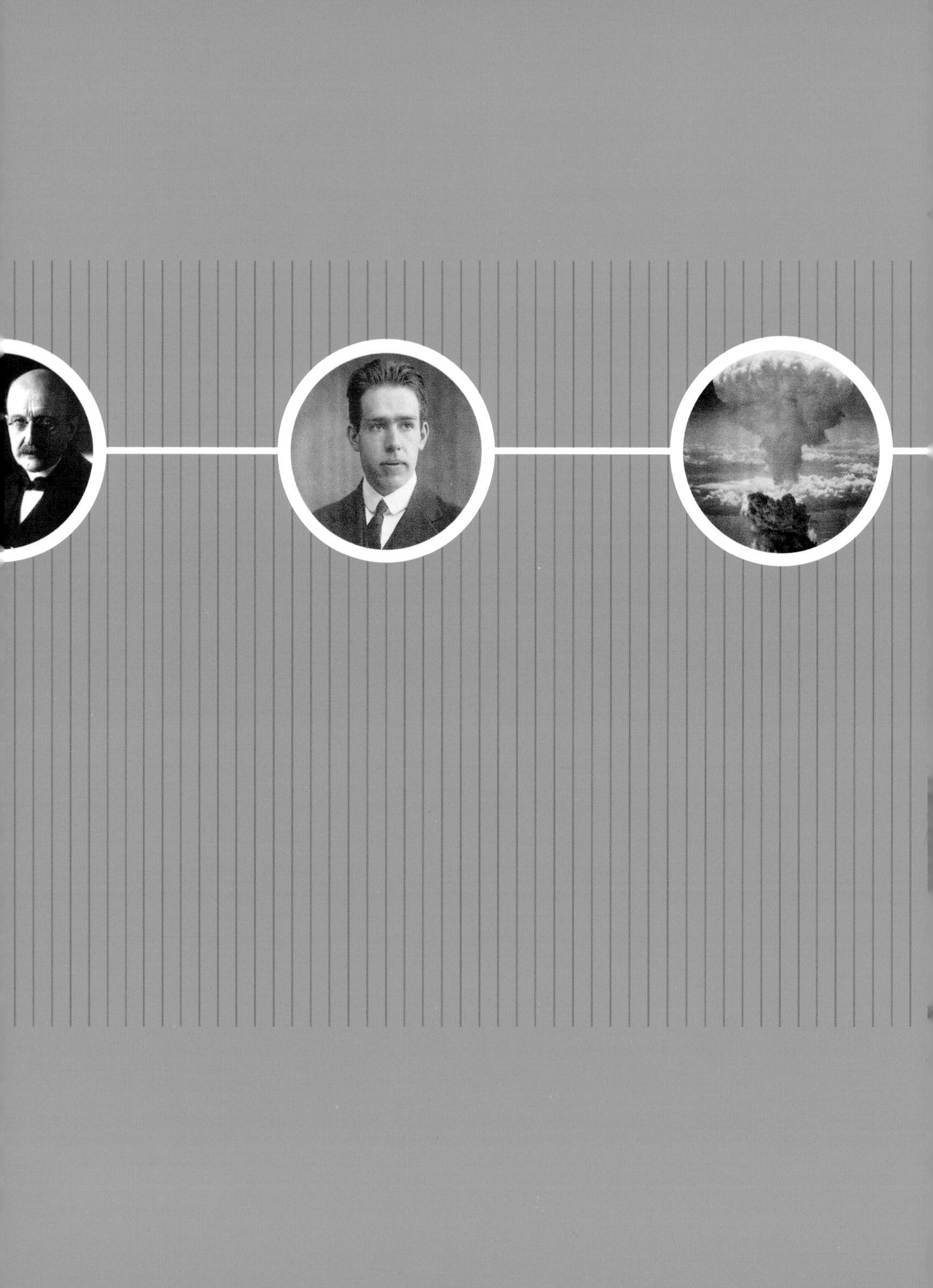

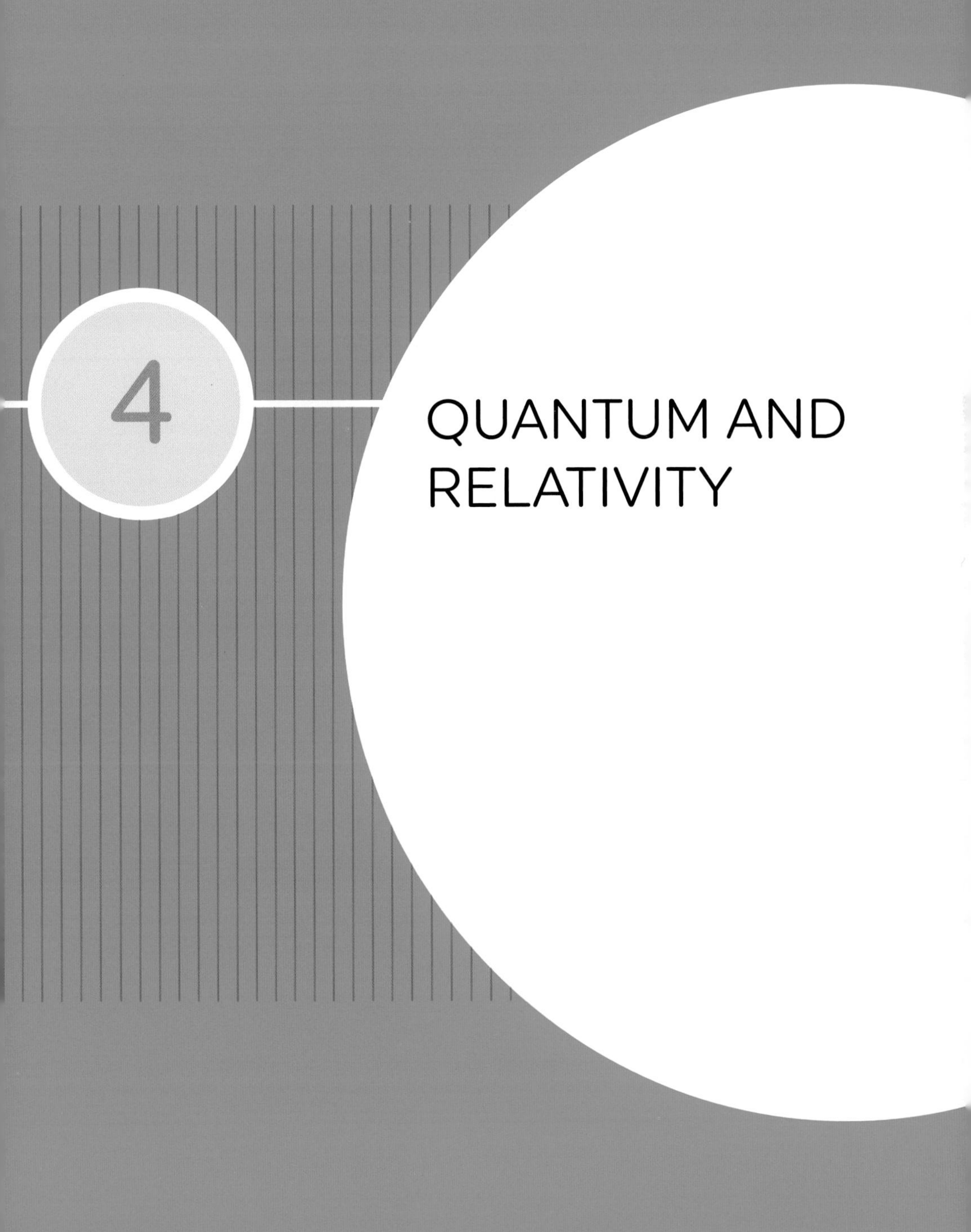

4 QUANTUM AND RELATIVITY

MAX PLANCK SOLVES THE ULTRAVIOLET CATASTROPHE

Classical physics worked well but not perfectly. Much like the heliocentric worldview that started to fall apart some 400 years before it, classical physics was starting to show its limitations. Something new was needed, and the first glimpse of this came in the form of Max Planck's solution to one of the biggest physics challenges in his time—the ultraviolet catastrophe.

Max Planck was born in 1858 into a middle-class and highly academic family. While he was still young, his family moved to Munich, where he attended the Maximilians school. He excelled and was pushed to achieve high grades, receiving special tutelage in mathematics, classical mechanics, and astronomy. He graduated early and began studying at the University of Munich, where initially he experimented with the diffusion of gases before undertaking a theoretical study of the world. He studied classical physics intensely, focusing on Maxwell's electrodynamics (see pages 96–99) and the laws of thermodynamics, and eventually writing his PhD thesis on the subject. From there he continued his work on thermodynamics, working as an unpaid private tutor while he sought a professorship, which he eventually got in Berlin in 1892. It was at Berlin University where he was tasked with looking at ways of making light bulbs more efficient. He quickly realized that his best approach to the problem would be to start with the theoretically perfect black bodies and work out the real-world complications after.

A black body is an imaginary theoretical object that absorbs all light and any other electromagnetic radiation that hits it. It also sends out all the light with 100% efficiency, meaning that if it had 100 joules of energy, it would emit all of it equally in all directions. While black bodies are an idealized scenario and most real world objects are gray bodies (ie, they have less than 100% efficiency in absorbing and emitting radiation), black bodies can be used as good approximations to real objects, such as stars, in order to test theories.

In 1905, British scientists Lord Rayleigh (1842–1919) and James Jeans (1877–1946) presented the Rayleigh–Jeans law, which describes how much radiance (the amount of radiation being given out from a set amount of surface area) is emitted by a black body at any chosen wavelength. The law works perfectly, but only for long wavelengths.

TRANSFORMATIVE WORK: A photograph of Max Planck from the early 1930s. His unexpected discovery revolutionized physics perhaps more than any other.

When the wavelengths start to reach 3000mm and below, which is just into the ultraviolet spectrum, things start to go very wrong. At this point, the Rayleigh–Jeans law predicts that all black bodies will radiate an infinite amount of energy. When applied to real world objects, this means that every single star, planet, and bright object should produce enormous quantities of energy in the ultraviolet wavelengths. This simply doesn't match up with reality, as we can see in the graph below. This problem became known as the ultraviolet catastrophe.

PLANCK'S SOLUTION

The ultraviolet catastrophe was already well known by the time Planck set about attempting to make more efficient light bulbs. A few others had tried to solve the problem, most notably Wilhelm Wien (1864–1928), who in 1896 proposed an solution, known subsequently as Wien's law, that solved the

THEORETICAL AND EXPERIMENTAL RESULTS:
A graph showing the differences between the theoretical prediction and the experimental result of the Rayleigh-Jeans Law.

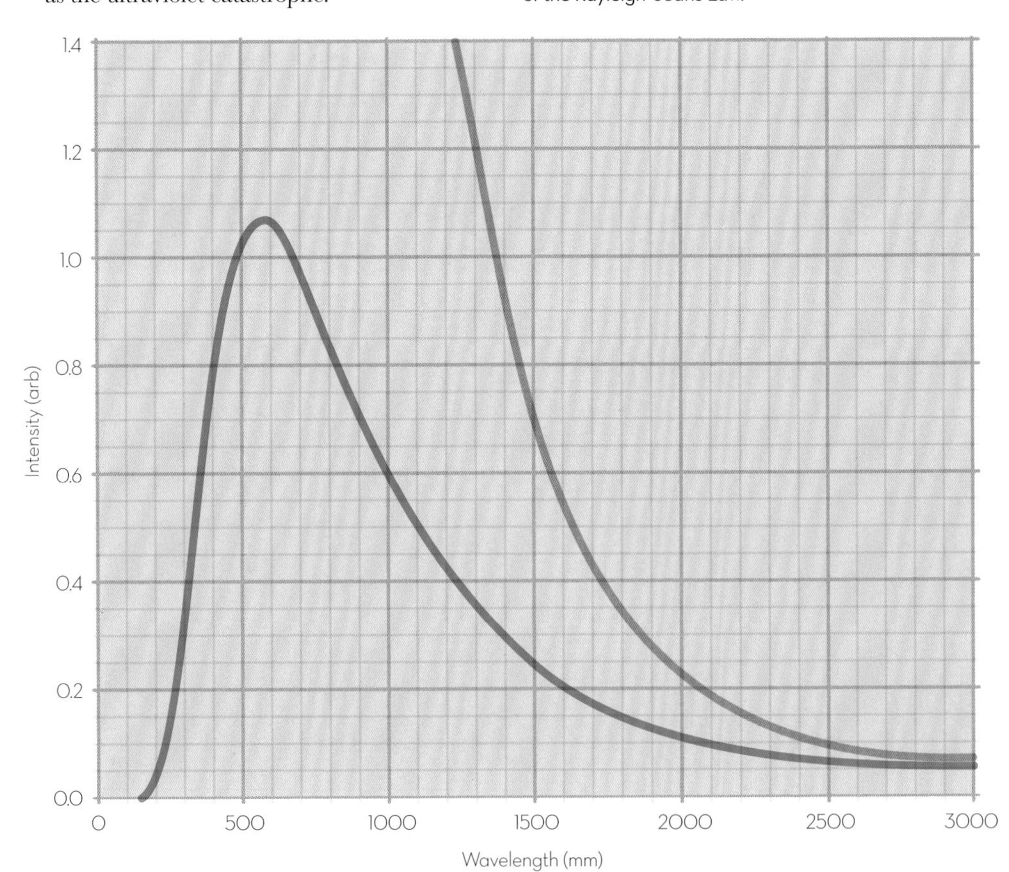

problem for small wavelengths but fell down at the larger wavelengths. Planck did some work on deriving Wien's law and attempted a few modifications, but he was unable to make the theory match the observations.

At this point, Planck made a bold and ultimately profound assumption, which is known today as the Planck postulate. It states that electromagnetic waves can be produced only in quantized (set) amounts. This is to say that energy can only come in certain values. With this insight, Planck went on to devise a formula, based around a mathematical constant that he had created, known today as Planck's constant, which successfully combined the two incomplete earlier theories. It matched Wien's law for low wavelengths and also the Rayleigh–Jeans law for high wavelengths. And yet more importantly, it matched the observational data.

THE RELUCTANT REVOLUTIONARY

With his theory of quantization, Planck had prompted, albeit unwittingly, an incredible leap forward, heralding the arrival of quantum physics. It was possible to use his eponymous constant to derive new fundamental measurements, such as Planck time and Plank length, which are the smallest units it is feasibly possible to consider.

Planck had changed everything, yet when asked about it later he described his constant as a "purely formal assumption... actually, I did not think much about it." In fact, for many years his findings raised very little fuss among academics, and Planck himself tried very hard to make them compatible with classical physics.

ALL THAT REMAINS

Max Planck was famously told by Philipp von Jolly (1809-1884), a professor of physics at the University of Munich, that "in [physics] almost everything is already discovered. All that remains is to fill a few unimportant holes." Planck responded saying that he wasn't there to discover anything new, only to understand. It would seem they were both very wrong.

It wasn't until 1905 that the scientific community took note. Albert Einstein presented his solution to another problem that classical physics couldn't explain—the photoelectric effect—by use of Planck's quantized energy, thereby showing that the quantization of energy was real. Even then Planck protested, attempting instead to champion Maxwell's classical theories on electromagnetism. However, after Einstein showed in 1910 that certain anomalies relating to specific heat capacity, a measure of the amount of heat energy that's required for a given object to change temperature by 1°, could only be solved through quantum theory, Planck was finally convinced.

EINSTEIN'S *ANNUS MIRABILIS*

Albert Einstein (1879–1955) is now known as one of the most famous physicists of all time, and yet he started out as an average and somewhat unruly student. But in 1905, when stuck in what felt to him like a dead-end job, he produced four scientific papers that changed physics almost entirely.

Albert Einstein was born in 1879 in the small town of Ulm, Germany. In 1880 his family moved to Munich so that his father, an entrepreneurial engineer and salesman, could set up a company making electrical equipment. Einstein received an education through the public school system and was considered to be a largely unremarkable child. When he was fifteen his father's company lost out on a huge contract to supply Munich with lighting, prompting his father to move the family in order to find work.

While his family resettled in Pavia, Italy, Einstein initially remained in Munich to complete his studies, but after causing much fuss, and with the help of a doctor's note, he rejoined his family. He spent the next few years attempting to get into Swiss universities, even renouncing his German citizenship (though this was likely to avoid military conscription). He was an average student in most subjects, but he excelled in mathematics and physics, and at the age of seventeen he was able to enroll on a mathematics and physics teaching degree at Zurich Polytechnic.

After graduation he was unable to find a teaching post, so he took a job in a patent office in Bern, Switzerland, where he evaluated mechanical patents. He found the work dull, but he was left with a lot of time to think. During the course of his work there, he did come across some of the problems and limitations of modern technology, specifically those from electrical signals and synchronized time. It is this experience that is thought to have led to his formulations of brand new ideas. Throughout much of this time, he continued to publish papers and continue his own research, but it was in 1905 that he would make many of his great contributions to physics.

THE *ANNUS MIRABILIS*

Annus mirabilis, Latin for "wonderful year," is how Einstein came to describe 1905. He was 26 years old and had just been awarded his PhD from the Zurich Polytechnic, and he was still working in the patent office. But in the course of 1905 he published four groundbreaking papers. Taken individually, each of them would have revolutionized physics, but to produce all four in quick succession elevated him almost overnight to the status of a scientific star.

A LESSON IN GENIUS: Albert Einstein pictured in 1921, during a lecture in Vienna. He later moved to the United States.

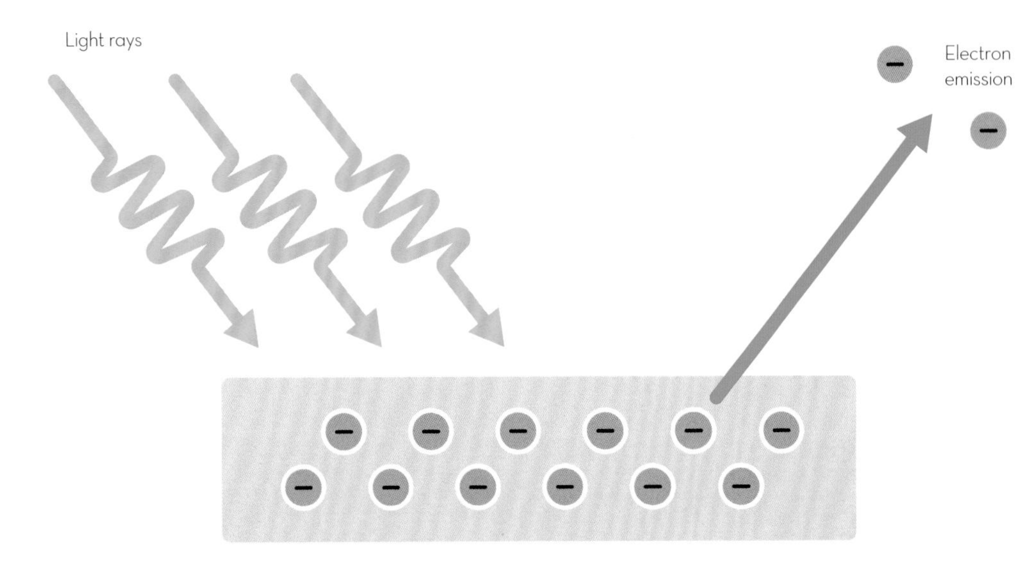

PHOTOELECTRIC EFFECT: A diagram showing how electron emission is caused from light via the photoelectric effect.

THE PHOTOELECTRIC EFFECT

The photoelectric effect is the process by which a piece of metal can be made to release electrons by shining a light on it. Experimentation established that the light needed to be of a certain frequency or higher for it to work (the exact frequency needed depends on the metal used). Yet this clashed with the theory, which stated that the light's frequency plays no part—to achieve the requisite energy for the photoelectric effect to occur was simply a matter of shining the light long enough.

Einstein resolved this opposition by describing light in terms of photons—wave-packets that deliver all the energy in a single burst. If the frequency isn't high enough, a given photon won't have enough energy to release the electron. With this he not only clarified the photoelectric effect but also showed that light has both wavelike and particlelike properties, thus starting the wave–particle duality theory (see pages 140–143). Furthermore, he showed that the photons would always have quantized amounts of energy, thereby raising Plank's quantum mechanics to the forefront of physics and giving it some solid experimental evidence.

BROWNIAN MOTION

In 1827 Robert Brown (1773–1858) observed particles trapped inside grains of pollen that were submerged in water. He noticed that these particles had a random, ever-changing motion that seemed to be caused by something, although he was unsure what.

Einstein was able not only to describe how the particles were moving but also to explain why.

His theory used statistical mechanics to describe the motion as being caused by innumerable tiny objects colliding with the particles, which eventually caused the motion. This was later confirmed experimentally by Jean Baptiste Perrin (1870–1942), and it became the leading proof for atomic theory—that something as small as a water molecule is composed of yet smaller parts.

SPECIAL RELATIVITY

Special relativity was a groundbreaking way of looking at the way in which objects in the universe interact. The theory postulated all sorts of strange phenomena, such as time dilation (the slowing of time) and mass increase.

MASS–ENERGY EQUIVALENCE

$E=mc^2$ is perhaps the most famous equation in all of physics. However, this is not how Einstein originally wrote it. In the 1905 paper the equation is written as $m=\frac{L}{c^2}$ (L is used instead of E). The reason for this is that the paper entitled "Ist die Trägheit eines Körpers von seinem Energieinhalt abhängig?" ("Does the Inertia of a Body Depend Upon Its Energy Content?") looks at an object's inertia (mass).

In the paper, Einstein uses the equation to explain the effect of energy on the mass of a system. This essentially means that if you had two identical pendulums, and one of them was made to start moving, the moving pendulum would have more mass than the stationary one. However, because c^2 is such a huge number (9×10^{16}), the additional mass is always incredibly small and pretty much unnoticeable in our daily lives.

In 1932 the mass of a neutron was first measured, and it became immediately apparent just how important Einstein's mass–energy equivalence was on a sub-atomic level. If you take a hydrogen atom, which is made of one proton and one electron, you find it weighs less than one proton and one electron on their own. The missing mass is converted into "binding energy," which is what keeps the atom together. It is the breaking and releasing of binding energy that makes nuclear fusion and fission possible.

THE GREATEST PHYSICIST OF ALL TIME?

In his lifetime Albert Einstein produced over 300 scientific papers. He revolutionized almost every aspect of our understanding of the universe, making huge contributions to physics in the subatomic realm and far beyond. All of this makes him potentially the greatest physicist of all time.

THE GEIGER–MARSDEN EXPERIMENT COMPLETES THE ATOM

The atom—a tiny fundamental thing that represents the basis for everything in the universe—had been an idea in physics since the ancient Greeks. However, despite the idea's age, very little was actually known about the atom. This all changed with the 1911 landmark publication of "The Scattering of α and β Particles by Matter and the Structure of the Atom."

Despite what the name suggests, the story of the Geiger–Marsden experiment is actually the story of three people: Hans Geiger (1882–1945), Ernest Marsden (1889–1970), and Ernest Rutherford (1871–1937). At the time of the experiment, Ernest Rutherford had already earned his place in history thanks to his vast amount of work on radiation. He had discovered the three types of radiation—Alpha (α), Beta (β), and Gamma (γ)—and had shown that they were intrinsically linked to atoms, especially in the way that atoms decay. Geiger had gained a doctorate in mathematics and physics from the University of Erlangen in Germany. He traveled to the UK in 1907 and impressed Rutherford so much that he was offered a research position alongside him at the University of Manchester. Marsden was an undergraduate student at the University of Manchester who joined Rutherford and Geiger's research group in 1909.

The team was working on experimenting with Rutherford's α-particles. They knew that these particles were created by certain elements, such as radium or uranium, and

GEIGER COUNTER A diagram of an early Geiger tube, used for detecting α particles. The screen "Z" will flash when it is hit by an α particle.

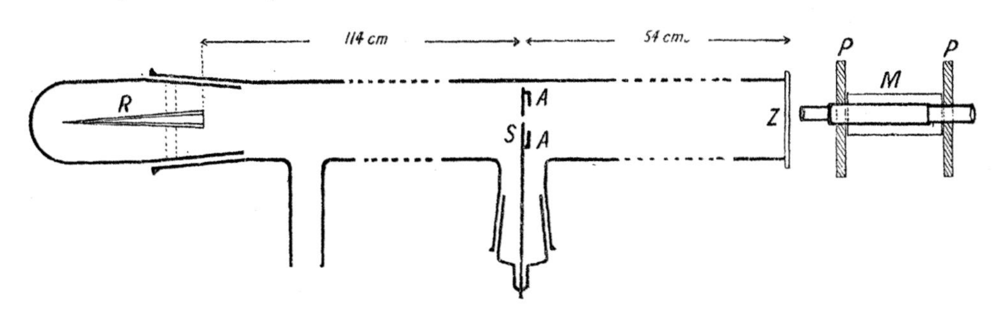

RUTHERFORD'S CHARGES: Hans Geiger (left) and Ernest Marsden (right). They were both students of Ernest Rutherford, who headed up the team for their quest for discovery.

that they were positively charged, but that was about it. Rutherford particularly wanted to know what their charge-to-mass ratio was. To do this he wanted to be able to count the number of α-particles that are released, while measuring their total charge. While α-particles are too small to be seen, they create a process called "ionization," which creates charged ions, which in turn can be used to create a current. Using this as a basis, Rutherford and Geiger created a tube that would ideally count the pulses of electricity as α-particles passed through it. However, the device failed to work because the particles would scatter erratically, with some producing more or fewer ionized events than others. This was confusing because it meant that there was more deflection of the α-particles than there should have been.

Rutherford and Geiger then developed a technique whereby a fluorescent screen gave off a tiny flash of light whenever an α-particle hit it. Using this, however, was tedious work that required spending hour after hour in a dark room hunched over a microscope counting tiny flashes. Unable to endure the surprising amount of effort it took to do this, Rutherford tasked Geiger with exploring the effects of scattering α-particles.

THE EXPERIMENTS OF 1908 AND 1909

In 1908 Geiger constructed a long glass tube with a source of α-particles at one end (in this case, radium, because it emits large amounts of α) and a slit barely 1mm in size at the center. The α-particles would pass down the tube and through the slit, causing a small glowing patch on a fluorescent screen at the opposite end of the tube. Geiger found that by removing the air from the tube, the patch of light would become more focused, and when he let air back in again it would spread out. He also discovered that by placing a very thin foil of gold (only several atoms thick) it would cause the light to spread out even further. This proved that α-particles were scattered by both air and solid matter.

In 1909 Marsden joined the team and worked with Geiger on looking at scattering at large angles (the previous experiment was limited to only a small range). They set up an α-source in front of a metal foil and then positioned the screen at various points along an arc, to see what would happen. They found that there were α-particles being deflected off at angles up to 90°. They also found that if the foil was made from heavier elements, such as gold, rather than lighter ones like zinc, there would be more scattering. The results of these experiments were published in "On the Scattering of α-Particles by Matter and On a Diffuse Reflections of the α-Particles." Inspired by their success, Geiger and Marsden built a better, more accurate experiment to examine the effect further.

RUTHERFORD EXPERIMENT: An illustration of the set-up for the Rutherford scattering experiment. The circular fluorescent screen meant that scattering at any angle could be detected, which was a huge improvement on the earlier Geiger tube.

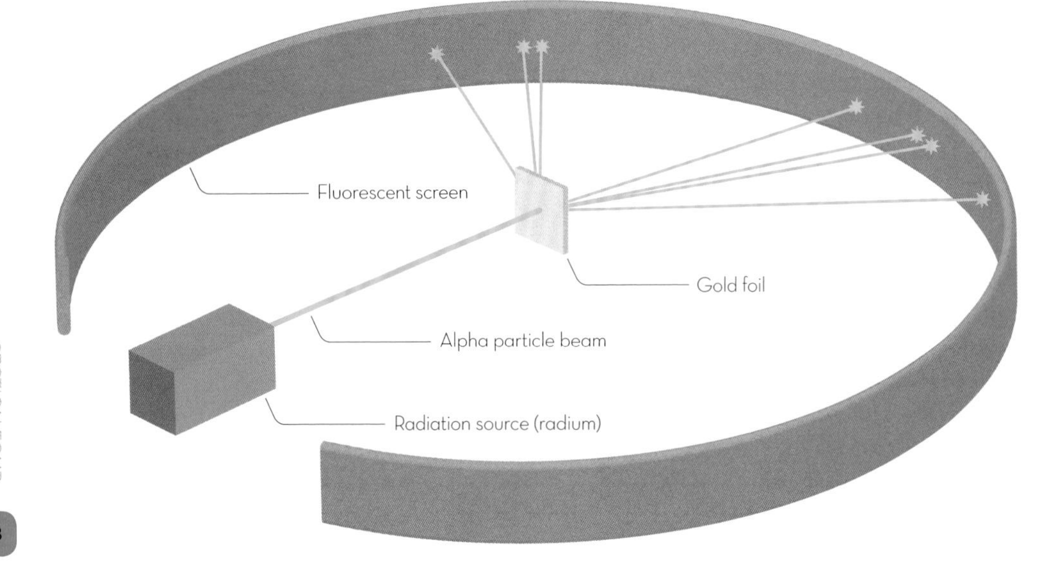

THE "PLUM PUDDING" MODEL

The main model for the structure of the atom before the Geiger-Marsden experiment was the model created by J. J. Thomson (1856-1940), which is often called the plum pudding model. It consisted of a large positively charged nucleus that was dotted with small negatively charged particles called electrons (which Thompson had discovered earlier), in much the same way that a plum pudding is filled with smaller plums.

THE GEIGER–MARSDEN EXPERIMENT

The seminal experiment was performed in 1910 and consisted, much like the previous experiment, of an α-source fired at a thin sheet of gold, with the resulting scattered particles detected by a fluorescent screen. This time the experiment was sensitive enough for each individual particle to be counted at all angles by use of a circular screen (see the diagram, left). From this Rutherford was able to calculate a formula that tells us how many events we expect to see at any given angle:

$$N(\Theta) = \frac{N_i n L Z^2 k^2 e^4}{r^2 E_k^2 \sin(\Theta/2)}$$

One of the most striking aspects of the findings was the realization that, if they placed a detector directly behind the beam, they would occasionally find an event! Rutherford said this about it:

> *It was quite the most incredible event that has ever happened to me in my life. It was almost as incredible as if you fired a 15-inch shell at a piece of tissue paper and it came back and hit you.*

> *On consideration, I realized that this scattering backward must be the result of a single collision, and when I made calculations I saw that it was impossible to get anything of that order of magnitude unless you took a system in which the greater part of the mass of the atom was concentrated in a minute nucleus. It was then that I had the idea of an atom with a minute massive centre, carrying a charge.*

Prompted by this discovery, Rutherford developed an entirely new model of the atom, known today as the Rutherford model. It has a dense but small central nucleus that has the same charge as the α-particle, which causes the occasional complete backscatter, lots of empty space, which results in most of the particles getting through unscattered, and the occasional electron, with an opposite charge to deflect the α-particles. The Rutherford model is largely the same as we use today (there are a few small differences), and its discovery opened the way for a huge amount of atomic research, both in physics and in chemistry.

NIELS BOHR EXPLAINS SPECTRAL LINES

Although Ernest Rutherford had established an accurate model of the atom, there remained many things to explain. Foremost among these was the phenomenon of spectral lines, the mysterious lines of light produced by any given element or chemical. It would take the genius of Niels Bohr (1885-1962), and more than a little quantum, to solve it.

Niels Bohr was born in Copenhagen, Denmark, to a moderately affluent middle-class family. He had a good childhood and became a student of physics at Copenhagen University in 1903. Although it was a small department, consisting of a single lecturer, he quickly developed a love for the subject. In 1905 Bohr turned his efforts to a challenge, presented by the Royal Danish Academy of Sciences and Letters, to determine a method of measuring the surface tension of liquids using a system of vibrations put forward in 1879 by the eminent Lord Rayleigh (1842–1919). Using equipment borrowed from his father's laboratory (the university did not have its own dedicated physics laboratory), Bohr managed not only to demonstrate how it could be done, but also to improve the method, and duly won the prize. In 1911 he wrote his thesis on electron theory, which showed that it was unable to fully explain magnetism. It was a groundbreaking paper, but it received little publicity, probably due to it having been written in Danish.

In 1911 Bohr traveled to England, where he met J. J. Thompson (of "plum pudding" model fame). After failing to impress him Bohr was invited to work with Rutherford on his new model of the atom. After a year

ATOMIC PHYSICIST: A photograph of Niels Bohr as a young man. His work with atomic structure led to him becoming a key player in the Manhattan Project.

working with Rutherford, Bohr returned to Denmark to marry and was appointed to a professorship at the University of Copenhagen, where he taught thermodynamics. It was while working there that his three famous papers, known collectively as "the trilogy," were published. They set out the Bohr model of the atom and also explained how spectral lines form.

WHAT ARE SPECTRAL LINES?

Spectral lines had been discovered in 1885 by Johann Balmer (1825–1898) while he was working with hydrogen. He had discovered that upon exciting hydrogen gas (by passing electricity through it) and then separating out the resulting light it gave off, a series of set lines were created. These spectral lines would always have the exact same wavelengths as given by the following formula:

$$\lambda = \frac{1}{R_H\left(\frac{1}{2^2} - \frac{1}{n^2}\right)}$$

Where λ is the wavelength, R_H is the Rydberg hydrogen constant, and the value of n changes for each line. The lines he found became known as the Balmer series. While these had been well documented since their discovery in 1802, and were found for a large number of different elements and molecules, a definitive explanation for them eluded scientists.

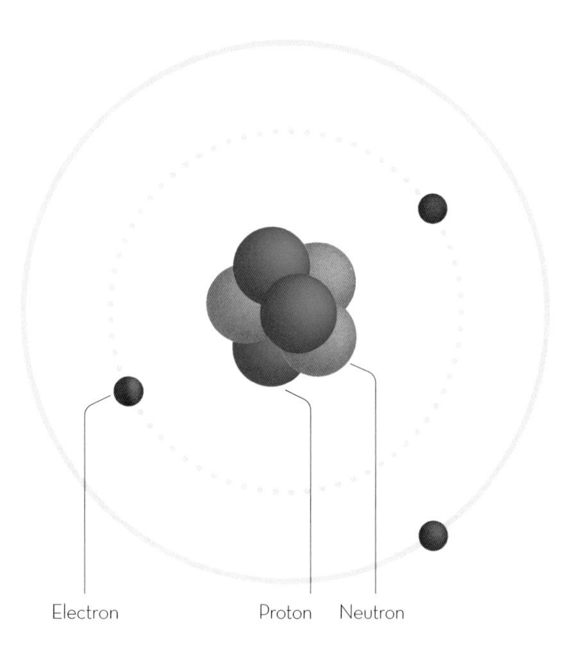

PLANETARY MODEL: An image of the Bohr model of the atom. It is often called the "planetary model," due to the similarity of its layout to that of the solar system.

THE BOHR MODEL

The diagram of the Bohr model is an image that we are all familiar with, with the central nucleus orbited by electrons, very much like a miniature solar system. This orbital system was created not by Bohr but by Hantaro Nagaoka (1865–1950), in 1904. However, it was Bohr who managed to modify it to explain how spectral lines were being formed.

Bohr started with the standard Rutherford model of the nucleus, with electrons orbiting it, but he realized that if this were the case, atoms simply couldn't exist. Classical mechanics predicts that as an electron orbits

THE USES OF SPECTROSCOPY

As we previously noted, every chemical gives off different, constant spectral lines. But what makes it so important is that every single chemical, element, and compound that exists gives off its own unique set of lines. No two chemicals ever look the same, and that's why spectral lines have been called chemical fingerprints.

This gives us a way to identify any chemical. You often find spectroscopy kits in laboratories or even places like crime scenes. Using a small sample, the machine is able to produce a spectrum and automatically read it to identify which chemicals are present.

But spectroscopy is even more incredible than that, because we know that the spectral lines for a chemical are universal. So it doesn't matter if the object is a bottle of a chemical mixture right next to us or a supernova right across the galaxy, spectroscopy still works. This means that by gathering the light from stars we are able to identify what they're made of. Recent advances in spectroscopy mean that by examining the light that has passed through the atmospheres of exoplanets (planets orbiting other stars), we are able to learn lots about them.

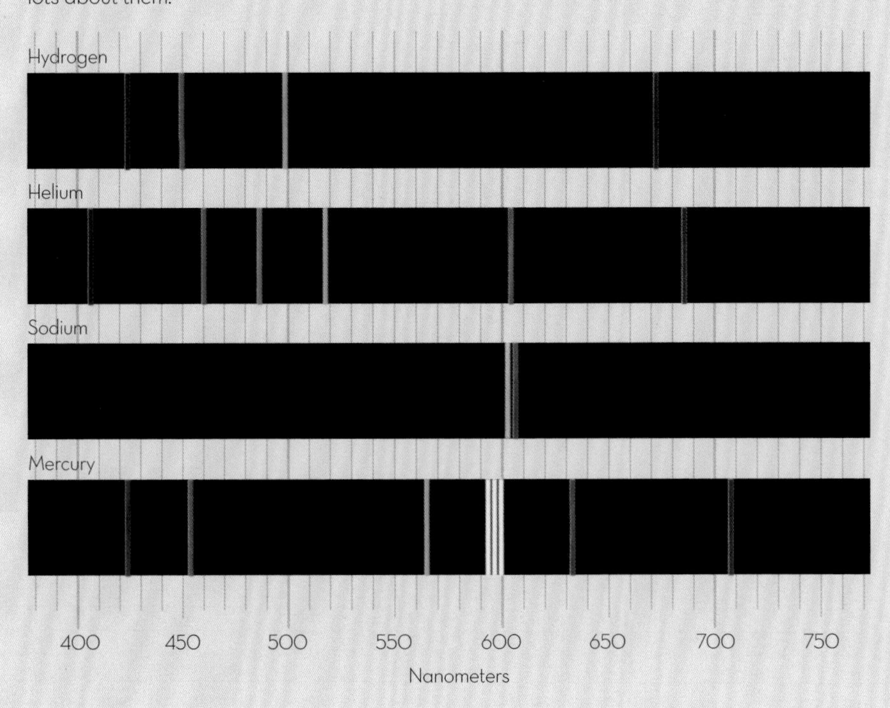

it would emit electromagnetic radiation (much like the spectral lines, only over all wavelengths, not just a select few). However, this would cause the electron to lose energy and, as this happened, the electron would slow down and spiral toward the nucleus due to the charge attraction. This would ultimately lead to the collapse of the atom in just 1×10^{-12} seconds. As this is clearly not the case, Bohr proposed three rules that would not only prevent the collapsing atom but also explain the spectral line emission:

- Electrons in an atom orbit the nucleus.
- Electrons can only orbit in certain "stationary orbits," which are at set distances from the nucleus. When in a stationary orbit an electron does not emit energy.
- Electrons can only lose or gain energy by moving from one stationary orbit to another.

This model fully explained the spectral lines, because the jumping from each energy level would always produce the same wavelength of light, and Bohr's calculations matched the observed results. This idea of stationary orbits is decidedly quantum in its nature, because it only allows set values, and the resulting energies are always some multiple of h, the Planck constant.

Despite its success, the Bohr model is far from perfect. Its equations only really work for a very limited range of simple atoms, such as hydrogen and ionized helium, because it can only describe a single electron. Equally, it assumes that electrons move only on a two-dimensional plane, which is not the case. There are also two issues with spectral lines that cannot be explained by the Bohr model:

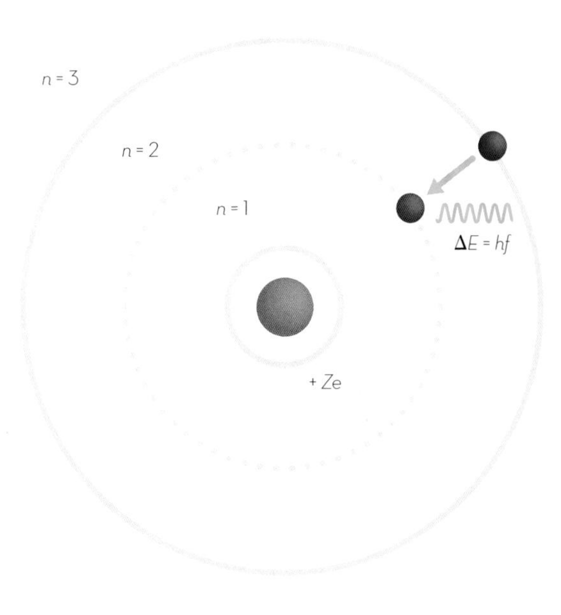

ENERGY AND TRANSMISSIONS: A representation of how energy levels and transmissions work in the Bohr model.

the fact that the different spectral lines have different brightnesses, and that with sensitive-enough equipment we find that the spectral lines can be split using a magnetic field (known as Zeeman splitting). The Bohr model was a good start that introduced the world to quantization in atoms, but it quickly became clear it wasn't the full picture.

EINSTEIN DESCRIBES GENERAL RELATIVITY

In 1915 Einstein published a series of four papers in which he described his new theory of general relativity. It was his magnum opus, and it would revolutionize physics in the biggest rewriting of the textbooks since Isaac Newton's *Principia*.

To put it in very simple terms, general relativity is a theory of gravity. It melds space and time together into space–time, which, although sounding like jargon, allows for the two to interact in very important ways.

In order to think about general relativity, we need to simplify the universe. Imagine it as a flat, stretched-out, two-dimensional piece of material that is flexible like Lycra. If you place a heavy ball onto the material, the material will stretch around the ball as it depresses the material out into three dimensions. Similarly, mass in our universe will cause space–time to warp around it in four dimensions. Gravity comes from other objects moving across the fabric of space–time, which then reach the dip caused by a large body and "fall" into it and move toward the larger body. In our analogy, this would be equivalent to rolling a marble across the material and watching it spiral toward the heavier object.

General relativity quickly became the accepted form of gravity, because it provides an explanation of all the things described by Newtonian mechanics, plus a number of other phenomena, such as issues with the orbit of Mercury and the redshifting of light by gravity.

"LIGHT'S ALL ASKEW IN THE HEAVENS"

The headline of the *New York Times* one day in June 1919 read "Lights All Askew in the Heavens." The first great prediction of Einstein's theory of relativity had come to pass. Arthur Eddington (1882–1944), a British astronomer, had traveled to the island of Príncipe off the west coast of Africa to observe the stars near the sun during an eclipse. He found, as Einstein had predicted, that stars that should not have been visible were there and that all their positions had apparently changed.

This deviation is caused by the distortion of light's path through space–time by massive bodies, such as our sun, which deflects the light around it. The light from stars is affected in this way, thus changing the apparent positions of the stars in the sky. Despite the roughness and somewhat spurious results of this first experiment, it was hailed around the world as solid proof of Einstein's theory. Later on, more accurate experiments using radio waves provided further confirmation.

It wasn't just the deflection of light that was predicted by general relativity. Other predictions that have since been proven correct include gravitational lensing,

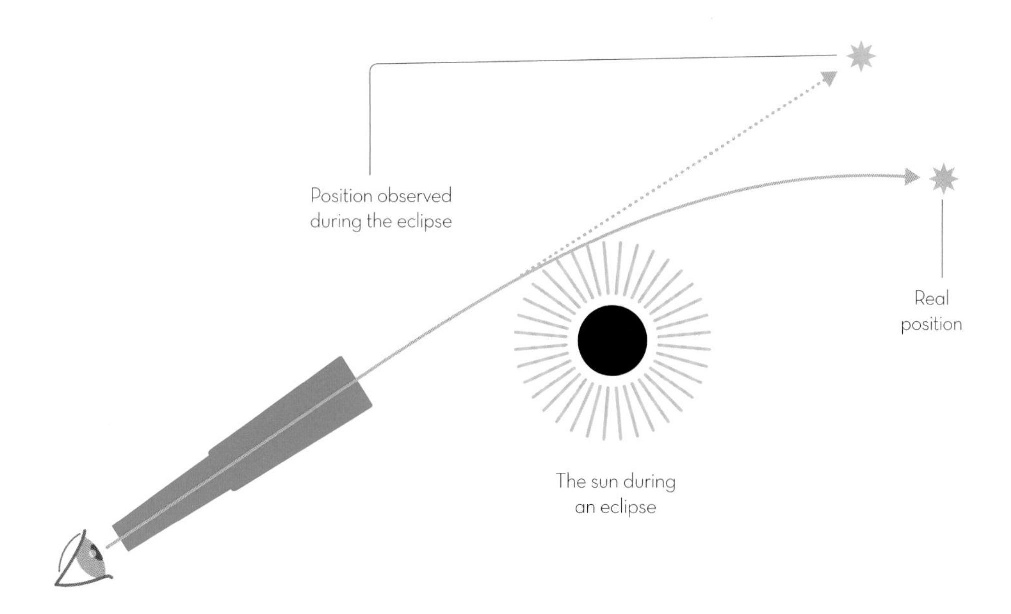

DEFLECTING LIGHT: As Arthur Eddington discovered during an eclipse, the sun deflects light, making stars visible that would otherwise be obscured.

gravitational time-delay, black holes, and gravitational waves (see pages 178–181).

SPECIAL RELATIVITY

Special relativity is a special case of general relativity. It is one in which all the objects we are observing are either at rest or moving at a constant speed. We start with two premises: The first is that the speed of light is always constant; the second is that there is no such thing as a preferred observer—one is as valid as any other. These are both simple and fairly understandable, but they have some strange consequences.

Returning to the example used earlier in the book (see page 103), we have A, B, and C all traveling at different speeds (0, 20, and 50 km/h respectively) and we want to know what their "real" speeds are. The existence of no preferred reference frame tells us that the speed of each person is entirely dependent on who is observing them. If you asked person A, they would say that they aren't moving, that B is moving at 20km/h, and that C is moving at 50km/h. But if you ask B, they would say that A is moving at 30km/h in the opposite direction to C, who is traveling at 20km/h. Both A and B are correct, because everything is relative.

While this may be difficult to comprehend, it is manageable. The answer to how fast A, B, and C are traveling depends on who you ask, and no one answer is more correct than any other. What makes the concept complex, however, is light—no matter who you talk to,

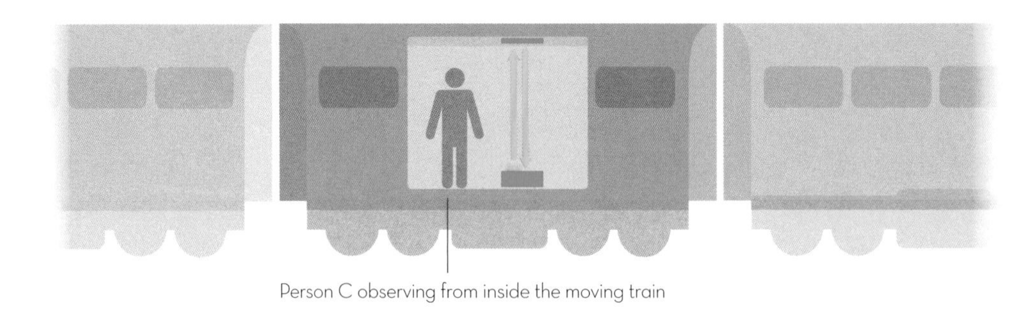

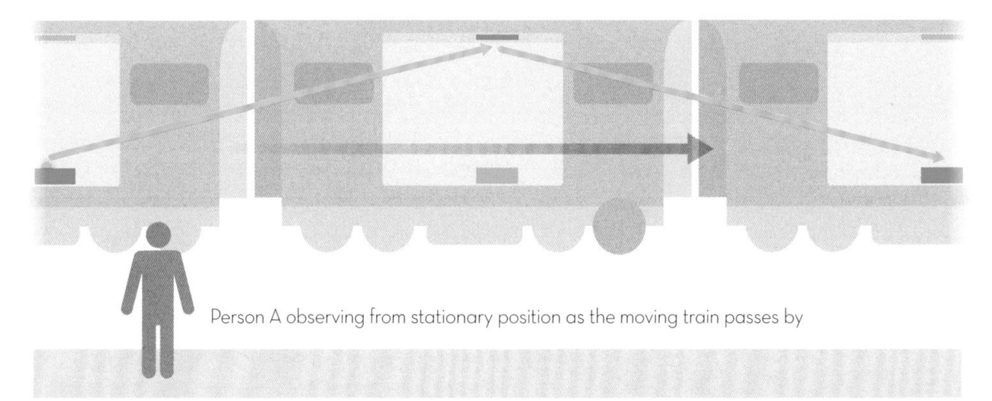

SLOWING TIME: An image showing how there is length change caused by motion. The speed of the light (shown here by the yellow arrows) remains constant, which means that time must slow.

they will always agree that a beam of light from their reference frame moves at *c* (3×10^8 meters per second), and this can cause some issues. Imagine that person C shoots a beam of light from the floor of the train carriage up to the top, where it hits a mirror and bounces back down again. Person C sees it travel up and down in a straight line at speed *c*. However, if A is somehow able to observe C do this, A would see the light move up and back down again but, due to the motion of the train, they will observe the train move along as the light travels, so its path up and down will form a triangle. Because the height of the carriage hasn't changed, they observe that the light has taken a longer path than C observed. They both agree that the light traveled at exactly the same speed, *c*. How is this possible? The answer is, the faster you move, the slower time becomes.

It's not just time either. As an object moves faster, special relativity shows that it will get shorter and heavier. So how have we never seen this happen before? Because it happens at incredibly high speeds—it only starts to get noticeable at about 70% the speed of light, which explains why we have never experienced it. We are, however, able to observe its effects in things like the subatomic particles "muons," which form in the outer atmosphere. They should decay before they reach the ground, yet don't, because they are traveling at near the speed of light. This means their time slows down, so they are able to travel a much farther distance before they decay away.

GENERAL RELATIVITY IN MATHEMATICS

Einstein created a mathematical basis for his special relativity, which he was then able to turn into general relativity. However, the mathematics for general relativity get exponentially harder, with even the "simple" equations involving huge amounts of work. So much so that the most complicated complete system to ever be fully explained with mathematics is the "Schwarzchild universe," which contains only a single particle of mass. In order to describe general relativity mathematically, Einstein had to treat the universe as a giant field, similar to an electrical or magnetic field. Once he had done this, he was able to produce the Einstein field equation:

$$R_{\mu\nu} - \frac{1}{2} R g_{\mu\nu} + \Lambda g_{\mu\nu} = \frac{8\pi G}{c^4} T_{\mu\nu}$$

This equation may not look too complicated, but don't be fooled. Each part of it is a special mathematical construct called a symmetric tensor, and they are four-dimensional to account for both space and time. This means that each section in their fully expanded form relies on ten individual components. To properly analyze a situation requires 40 simultaneous nonlinear partial differential equations. As you may imagine, this is not a trivial thing to do, and the complexity explains why only very simplistic universes have been described thus far.

Λ is the "cosmological constant." Einstein introduced it in 1917 in order to "hold back gravity," so that it was possible to make the universe neither expand nor contract. It was a convenience that he invented to try to sort out what he saw as a problem with his equations. He later described it as his biggest mistake, after Hubble proved that the universe is expanding. However, after experiments in 1998, it was discovered that Λ is in fact the exact value for the energy density of the vacuum of space. It would seem that even when Einstein gets something wrong he still manages to develop brand new physics ideas.

HEISENBERG INTRODUCES HIS UNCERTAINTY PRINCIPLE

Perhaps more important than learning new things in physics is discovering the limits of our knowledge, namely that there are some things that we may simply never be able to know. Werner Heisenberg's eponymous uncertainty principle goes beyond this, showing us what we cannot know and then making predictions based on it.

We all know that everything happens by cause and effect. If I hit a pool ball perfectly into a corner of the table, it will go straight into the pocket. If we set up the exact same shot a million times and everything happens in exactly the same way, the ball will always go into that pocket. The same is theoretically true for any shot you take in pool; given all the variables, a computer could calculate exactly where all of the balls will go when you take your shot. Surely then our entire universe is just the same? Sure, the universe is a lot larger than a pool table, containing around 4×10^{79} atoms and four different fundamental forces—electromagnetism, gravity, strong nuclear force, and weak nuclear force. But surely if we had a big enough computer, we would be able to calculate everything (theoretically at least—scaling up the best computer we have today to the size of the planet, it would take many times the lifetime of the universe just to calculate one room full of atoms, so such a device would not be practical). The Earth forming, the dinosaurs, the founding of America, and even you reading this book right now! It's a potentially scary idea and raises a lot of uncomfortable questions, but that's for the philosophers to worry about. As physicists, we remain largely unconcerned by its ramifications.

Quantum does not agree with this idea. One of the key points made by quantum mechanics is that we can never truly tell what is going to happen. We can make predictions based upon what is most likely, but in the end, we can't know with absolute certainty. So with my previous example of a pool ball, I might calculate that I have a 90% chance of getting it into the pocket. If we were to set up the same shot a million times, we'd see that nine out of ten times it would pot the ball, but one time out of the ten it wouldn't, even though everything down to the last atom is the same. While it may come as a relief that your entire life isn't controlled by equations and that you probably do have some free will, it can be difficult to understand how and why this happens.

NUCLEAR PHYSICIST: Werner Heisenberg, pictured in his office. He became an important figure in nuclear physics and much of quantum physics.

WHAT IS UNCERTAINTY?

Imagine you want to measure a piece of string, so you take a ruler and measure it to be 133mm (5¼in). Is that really how long it is? Probably not, because it's unlikely that the string ended exactly at the 133mm mark—instead, you rounded its length up or down. In scientific papers, this would be written as 133±0.5mm, to show that we've rounded to the closest value we can actually measure, so ΔL=0.5mm. Now let's say we were to use a more accurate laser device to measure it, we calculate the length, and we get 134,948,573nm. Even this won't be exactly correct; it will have been rounded up or down, with the actual value 134,948,573±0.5nm, with ΔL=0.5nm. So now that we know what uncertainty is, just how far can you go with it?

THE UNCERTAINTY PRINCIPLE

Heisenberg presented his uncertainty principle as:

$$\Delta x \Delta p \geq \frac{h}{4\pi}$$

This equation states that the amount of uncertainty in the position of a particle multiplied by the uncertainty in its momentum must always be larger than h (Planck's constant) divided by 4π (this equation is sometimes shown as $\hbar/2$ where $\hbar$ is equal to $h/2\pi$). The value of $h/4\pi$ is very small (5.27×10^{-35}), so the effects of this equation are not seen in day-to-day life. However, at the subatomic level it can become very important. If we reduced the pool ball to the size of an electron and had it moving around in a box, we wouldn't know exactly how fast it's moving or where exactly it is. If we did, then either Δx or Δp would be 0, which would break the equation. This also means that when we measure the position of the electron, the more accurately we do it (and hence the smaller Δx becomes), the less accurate the measurement of momentum becomes (as Δp has to get larger to keep it over the value of $h/4\pi$).

While you might understand the basic mathematics, it probably seems like not much more than a rather far-fetched idea. How do we see its effects in the world? Well, it turns out that people had been studying one of its

ENERGY–TIME UNCERTAINTY

There is more than one uncertainty principle in physics; another relates to energy and time. It looks like this:

$$\Delta E \Delta t \geq \frac{h}{4\pi}$$

This works in exactly the same way as Heisenberg's principle, only with different variables. The astonishing thing about this equation, though, is that it allows for the creation of particles from nothing! According to $E=mc^2$ (see page 115), a particle can pop into existence, so long as it is only for a very small amount of time. These are often called vacuum particles.

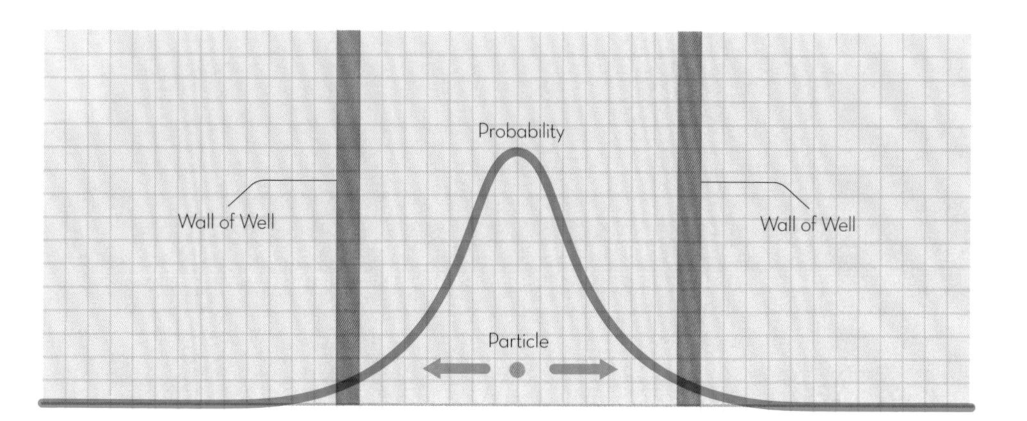

effects for around 20 years, α-decay. This is where an α-particle consisting of two protons and two neutrons spontaneously leaves the nucleus of a heavy atom (such as uranium). It was a well-documented but poorly understood process. The uncertainty principle can explain its occurrence through a process known as quantum tunneling. If we imagine a ball moving from side to side at the bottom of a well (see the diagram above), we use quantum mechanics to draw a "probability curve" (shown above in red), which tells us where the ball is likely to be and how likely it is to be there when we actually look at it. (Remember: because of the uncertainty principle, we can't calculate where it will be exactly.) The higher the curve, the more likely it is to be in a given place when we go to measure it. If you look at the curve, you'll notice that there is a small probability that the electron can be outside the walls. In classical physics, this is impossible due to the barrier, but quantum mechanics allows and predicts it.

PROBABILITY DISTRIBUTION: This diagram shows the probability distribution of a particle trapped in a well.

Returning to the problem of α-decay, it is evident that this is exactly what happens. The α-particle is contained within the nucleus, which holds it in using an "energy well," which can be drawn like the well above. The α-particle has a finite chance of being outside of the nucleus, and sometimes, spontaneously, this occurs, and the α-particle is emitted from the nucleus.

Heisenberg's uncertainty principle gave quantum physics the mathematical basis it needed to really get going. It also allowed for a much deeper understanding of some of the processes that can occur due to quantum effects. For its discovery, he was awarded the 1932 Nobel Prize in Physics, "for the creation of quantum mechanics."

EDWIN HUBBLE DISCOVERS THAT THE UNIVERSE IS EXPANDING

John Goodricke (see pages 68–72) extended our idea of how big the universe might be, and Edwin Hubble (1889–1953) made it even bigger. He not only showed there were galaxies beyond our own but also demonstrated that the universe itself is still getting bigger.

Edwin Hubble was born in Missouri and studied law at the universities of Chicago and Oxford, before returning home after the death of his father. He volunteered for the military in 1918, but the war ended before he saw combat. After the war he returned to England, this time studying astronomy at Cambridge, and thereafter he was quickly offered a staff position at the Mount Wilson Observatory in California.

When Hubble arrived at the Mount Wilson Observatory, he began using the recently built Hooker telescope, which was the largest in the world. He began observing the skies, focusing on categorizing Cepheid variables (see pages 70–71) and nebulae.

HUBBLE: Edwin Hubble using the 100-inch Hooker telescope at the Mount Wilson Observatory, California, in 1937.

A NEW GALAXY

On the November 22, 1924, the *New York Times* published an

article entitled "Finds Spiral Nebulae are Stellar Systems," in which Hubble showed that the nebula M31 was in fact another galaxy. He published a more formal paper early the following year. With this discovery Hubble had opened the door to extragalactic astronomy and proved we're not the only galaxy around, opening our eyes to the fact that there is more out there to discover.

But he didn't stop there. Using set sources of light (such as Cepheid variables), he found that most distant objects in space were red-shifted, meaning that the light they gave off moved wavelength toward the red end of the visible spectrum before it reached us. Not only that, but objects that were further away showed more red-shift. It was realized that this red-shift was being caused by the Doppler effect, which causes waves from moving objects to shorten or lengthen depending on their movement.

What did the red-shift show us? It showed that almost every object in the universe is moving away from us. This could have been interpreted as evidence that Earth is at the center of the universe. Hubble, however, knew better than this. He posited instead that everything is moving away from everything else. In other words, the universe itself is expanding! It logically follows that, by running time in reverse, you would find that the universe emerged from a single point—an event known as the Big Bang. Hubble expansion became the leading proof of the Big Bang.

The question then is, if everything is expanding away from everything else, where is the center of the universe? The answer is complicated significantly by the fact that the universe is four-dimensional, whereas we experience it in three dimensions. The best way to think of it is to pretend that the universe is a three-dimensional balloon and that we experience it in two dimensions (i.e., the surface of the balloon). If you were to cover the surface in dots to represent the galaxies and then inflate the balloon, the dots would all move away from each other as the balloon expands. The actual origin is now in the center of the balloon and not reachable in our two-dimensional space.

CMB

The Hubble expansion also explains a phenomenon known as the cosmic microwave background (CMB). It was accidentally discovered by Arno Penzias (b. 1933) and Robert Wilson (b. 1936) as a microwave signal that can be heard anywhere in the sky. Drawing on Hubble's work, it was realized that this was a very red-shifted version of the first ever light, created in the universe in the "recombination event"—the point at which the universe had cooled enough to allow light to move around.

GÖDEL WRITES HIS INCOMPLETENESS THEOREM

As we have seen with Heisenberg's uncertainty principle, sometimes it is what we can't know that's more interesting in physics. And Kurt Gödel (1906–1978) shook the scientific community by demonstrating the limitations of mathematics.

Kurt Gödel was born in Brün, at that time part of Austria-Hungary. He was a star pupil and attended the University of Vienna. He studied theoretical physics but quickly developed a love of mathematics and philosophy, eventually specializing in mathematical logic, which he described as "a science prior to all others, which contains the ideas and principles underlying all sciences." He came across the book *Grundzüge der Theoretischen Logik* (*Principles of Mathematical Logic*) by Wilhelm Ackermann (1896–1962), in which a question was posed that piqued his interest: Are the axioms of a formal system sufficient to derive every statement that is true in all models of the system? In other words, can we prove everything?

After gaining his PhD Gödel remained in Vienna and published his theorem in his 1931 paper "On Formally Undecidable Propositions in Principia Mathematica and Related Systems I." It comprised two theorems:

FIRST INCOMPLETENESS THEOREM *Any consistent formal system F within which a certain amount of elementary arithmetic can be carried out is incomplete; i.e., there are statements of the language of F which can be neither proved nor disproved in F.*
EXPLANATION If the system is consistent, it cannot be complete.

SECOND INCOMPLETENESS THEOREM
Assume F is a consistent formalized system that contains elementary arithmetic. Then F does not prove Cons(F).
EXPLANATION The consistency of the axioms cannot be proved within the system.

THE MEANING OF THE INCOMPLETENESS THEOREM

Gödel's theorem is not easy to read or understand, and its formal proof is frankly mindboggling, but the general meaning is fairly clear. Whenever we try to prove anything, we need to use some kind of system. The ones we use in physics are mathematics and logic. Underlying any system is a set of axioms—statements that we assume are true and on which the rest of the system is based. For example, in algebra we have a number of axioms, including:

The reflexive axiom $a = a$
Symmetric axiom *If* $a = b$*, then* $b = a$
Additive axiom *If* $a = b$ *and* $c = d$*, then* $a + c = b + d$

Although these may seem obvious, they are nonetheless assumptions. There is no real need for $a = a$ to be true, it is just our choice. The second of Gödel's theorems proves to us that there is no way for a system to prove its own axioms. This shows then that the axioms you choose will always be the limits of your system.

The first theorem tells us something far more unsettling. If the system we set up has rules that are always the same, there will always be unanswerable questions. Take for example this equation: $42 \div 0 = x$. It is properly formed, yet there is no answer to it. There are plenty more examples in mathematics of unanswerable questions, and in order to answer them a new form of mathematics is required (the imaginary number i is one such example) but again, even here Gödel's incompleteness theorem tells us that there will be questions for which there is no answer.

So where does it leave us? Ultimately, it tells us that, try as we might, our methods for exploring the universe are imperfect and will only ever produce imperfect results. Furthermore, there will always be questions to which we are unable to find the solutions. But that's never going to stop physicists from trying!

LATERAL THINKING: Kurt Gödel, photographed at the Institute of Advanced Study. His interest in philosophy as well as physics encouraged him to look at things in a different way.

FRITZ ZWICKY REALIZES MOST OF THE UNIVERSE IS MISSING

With the creation of yet more powerful telescopes and tools, the universe seemed to be revealing all of its secrets to astronomers. However, a rather unassuming calculation on a galactic cluster by Fritz Zwicky (1898–1974) revealed that we are actually missing most of it.

Fritz Zwicky was born in Varna, Bulgaria, where his father was the ambassador of Norway for the area. When Zwicky was six he moved to live with his grandparents in Switzerland, where he began his studies, eventually going on to study mathematics and physics at the Swiss Federal Polytechnic. At the age of 27 he was offered a research post at the California Institute of Technology, where he focused mainly on supernovas (a word that he himself coined for the phenomena of large stellar explosions).

In 1933 Zwicky was performing a number of calculations using the virial theorem—a system for looking at the energies of a stable system of a large number particles—on galaxy clusters. In particular he focused on the Coma cluster, which is a large grouping of over 1,000 galaxies some 321 million light-years away from Earth. The mass of the galaxies was measured using their brightness, as was standard practice. However, Zwicky calculated that the galaxies were moving far too fast and when he worked it backward from the velocity, he realized that the mass in the Coma cluster must be around 40 times greater than the mass gained from the calculation based on its luminosity. So he dubbed this invisible mass-producing stuff *dunkle materie*—dark matter.

Zwicky's theory wasn't widely accepted until the publication of *Rotational Properties of 21 Sc Galaxies with a Large Range of Luminosities and Radii from NGC 4605* by Vera Rubin (b. 1928), which showed that many problems in astronomy (principally those to do with galaxies and larger objects) were solved by the introduction of dark matter.

GALACTIC ROTATION CURVES

We can map out the distribution of mass across our own galaxy through observation, from the center, where there is a huge concentration of stars and black holes, outward, where the amount slowly drops off. From here it's possible to predict the motion of the galaxy and how it will rotate. But we are also able to use radio dishes to measure the actual speed of the galactic rotation, and the result is very different.

STARGAZER: Fritz Zwicky pictured on Palomar Mountain, California, in 1937. His concept of dark matter filled many gaps in our understanding.

WHAT IS DARK MATTER?

Very little is known about dark matter, but there are a number of theories about what it could be:

BLACK HOLES: This idea has largely been discounted, because, while they do not give off light, their effects are clear to see. If the dark matter was in fact black holes, we would almost certainly notice them.

MACHOS: Massive astrophysical compact halo objects (MACHOs) would be normal stuff that has a lot of mass but doesn't give off much light, such as planets, brown dwarves, and neutron stars. However, even the most generous estimate of their numbers doesn't come close to accounting for the missing mass.

WIMPS: Weakly interacting massive particles (WIMPs) are the most promising candidate. They would be particles that are very small and only interact with the rest of the universe through gravity and the weak force. This would make them very difficult to detect, even though the amount of them required would be incredibly high.

MODIFIED GRAVITY: Maybe Einstein just got gravity wrong and it isn't always a linear relationship. Maybe the equations are not as simple as we think and gravity gets stronger around or outside of masses. It's just that we haven't worked out the equations yet.

MASS IN OTHER DIMENSIONS: A slightly more outlandish idea is that gravity can work through dimensions, meaning that the dark matter is just matter sat in a different dimension (hence we can't see it), but there is still enough of it to cause significant gravitational effects.

DARK ENERGY

Dark matter was just the beginning of the "missing stuff" mystery that has been perplexing physicists. An even bigger problem comes from dark energy—a type of energy that fills space and is the cause of the expansion of the universe (see pages 132-133). We know next to nothing about it, yet it accounts for about 68.3% of our entire universe (with dark matter, 26.8%, and normal matter, 4.9%, making up the rest). We know even less about it than we do with dark matter.

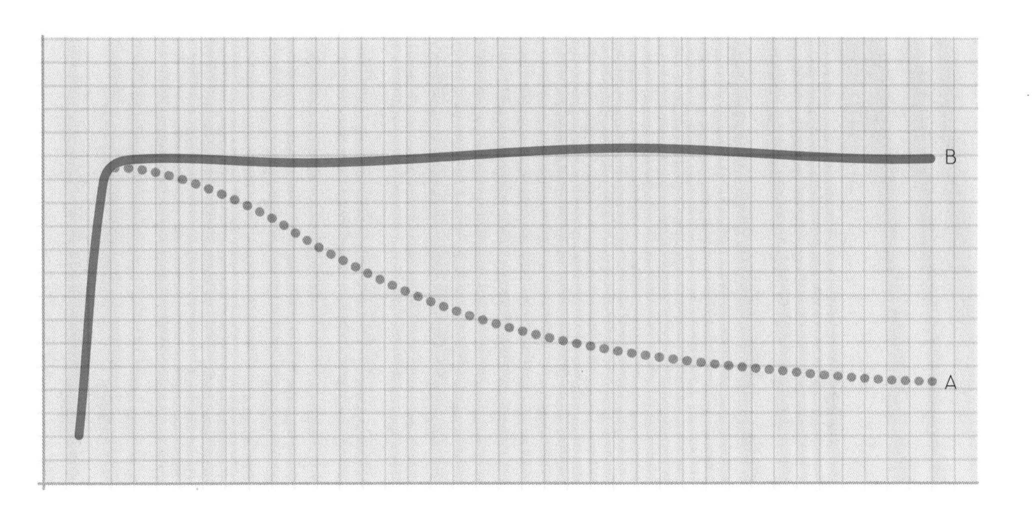

GALAXY ROTATION: A graph showing the differences between the theoretical prediction and the experimental evidence of the speed of the galaxy's rotation.

The expected rotation shown by line A is completely different from the real rotation shown by line B. The solution that Rubin put forward was that there is additional mass in the form of dark matter. The dark matter, curiously, is basically nonexistent at the center of the galaxy, and becomes more common the farther out in the galaxy you go, in order to make up for the missing mass. In all, about 90% of the mass in the galaxy is dark matter. This doesn't just happen in our galaxy either. Every galaxy we look at follows a similar pattern, and we find that there needs to be huge amounts of dark matter in all galaxies.

DARK MATTER OUTSIDE GALAXIES

It's not just in galactic rotation curves either that dark matter is needed. As Zwicky showed with the Coma cluster, large-scale bodies, such as galactic clusters and mega clusters, also need to contain dark matter, and huge amounts of it. The more normal mass there is, the higher the amount of dark matter there seems to be around it. From here it is possible to map out the dark matter, which we can find throughout the universe, into a huge cosmic web. Though what exactly it means and why it has formed like this are still unknown.

This missing matter is a huge problem; estimates suggest that there is around five and a half times as much dark matter as there is normal matter, also known as baryonic matter. To this day dark matter remains one of the largest knowledge gaps in astronomy and indeed all of physics. There are plenty of teams of scientists, hard at work attempting to identify dark matter in a number of different ways, who hope to be able to solve the mystery.

SCHRÖDINGER THINKS ABOUT CATS AND BOXES

Schrödinger's Cat is probably already familiar. It is a thought experiment created by Erwin Schrödinger (1887–1961) as a means of simplifying some very difficult quantum mechanics.

Erwin Schrödinger was born in Erdberg, Austria, to a deeply religious family of comfortable means. He received a relatively normal education and began studying physics at the University of Vienna. He was particularly interested in the philosophy of physics, likely due to his background, and began taking a more abstract and theoretical approach to his work. At the outbreak of the First World War he graduated from Vienna, and for the duration of the war he worked as an officer in the Austrian artillery. After the war, he took up a post at the University of Jena, in Germany, and in 1926 he moved to the University of Zurich.

It was at the University of Zurich where Schrödinger published his famous equation:

$$i\hbar \frac{\partial}{\partial t} \psi(r,t) = \hat{H}\psi(r,t)$$

It's a complicated equation that predicts the change of a quantum system over time. The most important part of the equation is the ψ function, which is known as the wave function. This represents everything about a system's quantum state (a system is any self-contained entity, such as an engine, a person, or a sealed box) and by solving it we can get an answer to any property of the system. We won't look at the equation here—it's incredibly complicated—but know that it can be used to produce incredible results.

QUANTIZATION The Schrödinger equation provides a rigorous mathematical basis for Planck's quantization, which can then be used to solve issues such as replacing the Bohr model of the atom.

WAVE–PARTICLE DUALITY The Schrödinger equation can be used to treat particles as a waveform, whereby, when placed under certain conditions, they exhibit wavelike qualities. Some claim that this proves all things are waves and simply show particlelike properties.

UNCERTAINTY The Schrödinger equation also gives a basis for Heisenberg's uncertainty principle within its wave functions, about which not everything can ever be known.

MATHEMATICAL BASIS: A photograph of Erwin Schrödinger from 1933. His formula was developed almost entirely mathematically, so it can be very difficult to map out its implications.

QUANTUM TUNNELING While Heisenberg had explained how quantum tunneling works, the Schrödinger equation mathematically proved it, and also predicted how often it will occur.

WAVE FUNCTION COLLAPSE The Schrödinger equation requires that ψ be answered in order to use it, so this means that to solve a problem using the Schrödinger equation we need to "collapse" ψ in order to gain a value for it.

WHERE DOES THE CAT COME INTO IT?

Quantum physics is confusing and mind-bending, and that's before we even start looking at the mathematics. Superposition of states, wave function collapse, wave–particle duality—none of it makes much sense. In his 1935 work *Die Gengenwärtige Situation in der Quantenmechanik* (*The Present Situation of Quantum Mechanics*) Schrödinger expressed the paradoxical nature of the system his equation had created through the now famous thought experiment, Schrödinger's Cat.

Imagine placing a cat in a box. The box is completely sealed and there is no way that you can tell anything about what's going on inside the box unless you open it. In this box is some sort of contrived device–the one Schrödinger described consisted of a Geiger counter next to a tiny radioactive source. Over the course of an hour there is a 50% chance that one atom of the radioactive material will decay, and there is a 50% chance that no atoms will decay. If it does decay, then the Geiger counter will tick, releasing a hammer to smash a vial of poison, which will then fill the box and kill the cat.

The important element here is that there is a random chance (which we have no way of predicting) that the cat inside the box is alive or dead. So which is it? In classical physics, the answer is quite simple—it is one or the other, but we just won't know until we open up the box. Quantum physics, however, says that the cat exists in a quantum superposition, whereby it is both alive AND dead until we open the box.

DEAD OR ALIVE? According to quantum physics, the Schrödinger cat exists in a quantum superposition, dead and alive until the box is opened.

Okay, so we have a cat that is alive and dead, which is a stand-in for any kind of quantum mechanical system for which there are multiple outcomes. Mathematically, the answer is undetermined until you solve the problem, which will not always give the same solution. While the problem might make some sense, the actual conclusion it reaches still seems very abstract. It can be hard to comprehend that if you find the cat had died some time ago, this somehow only happened when the box was opened. Unfortunately, this difficulty comes only with our ability as humans to comprehend it, because the math itself is perfectly sound. It may perhaps be a little easier to understand when we look at some of the potential interpretations as to how this works.

A QUANTUM HEADACHE

If you have still not really understood it, don't worry; even top quantum physicists struggle to understand how it might work in practice, choosing instead to focus on the mathematics. It has baffled the greatest scientific minds since its discovery, and it continues to do so today. Take Richard Feynman's word for it: "I think I can safely say that nobody understands quantum mechanics... Anyone who claims to understand quantum theory is either lying or crazy."

THE MANY WORLD INTERPRETATION Hugh Everett III (1930-1982) put forward a suggestion that in fact quantum mechanics is not something that tells us about how events occur but it informs us rather which path we take through them. It posits that all realities and outcomes are happening in alternate realities. When we collapse a wave function through observation, we simply move to a different reality. In the Schrödinger's Cat example, it is only when we open the box that it is decided which past occurred for us, thus explaining how the cat can have been dead for a period of time, even though the waveform only just collapsed.

OBJECTIVE COLLAPSE INTERPRETATION This is the idea that the cat (or any other system) is somehow observed by either itself or "the universe," meaning that wave functions collapse as soon as they form, making them a largely mathematical construct.

COPENHAGEN INTERPRETATION This is the form that is most commonly accepted as correct. At the point the box is opened, that waveform collapses and the cat is either dead or alive. This is to say, even when you find that the cat has died, it didn't actually die until the moment you opened the box, even if the cat had been "dead" for 30 minutes by the time you open the box. This is the one that perfectly follows the mathematics without any additional explanation, but it is equally the most difficult to understand.

TWO ATOMIC BOMBS ARE DROPPED ON JAPAN

The Second World War was unlike any other war ever fought. It was a technological war, conducted in laboratories and universities almost as much as on the battlefield. Communication technology, codebreaking, rocket development—all of these things contributed to the war, but the atomic bomb had arguably the biggest impact of all.

On August 6, 1945, a nuclear bomb was dropped on Hiroshima, Japan. Three days later a second bomb was dropped on Nagasaki. They were the culmination of around six years of work on the infamous Manhattan Project, in which a group of scientists based in the USA raced against their German counterparts in the Uranprojekt to produce the first usable nuclear bomb.

The Manhattan Project was led by J. Robert Oppenheimer (1904–1967). The scientists under his supervision had a great number of issues to contend with while attempting to build the bomb: developing suitable materials, calculating how best to construct it, and making sure that it would explode as intended. It was after the attack on Pearl Harbor, the point at which America entered the war, that

DEATH UNLEASHED: An image of the mushroom cloud caused by the atomic bomb dropped on Nagasaki on August 9, 1945.

the project really got going. The size of its staff was increased enormously and leading American, Canadian, and British scientists were brought in from fields as diverse as thermodynamics, electromagnetism, and condensed matter.

The first test of a nuclear weapon was codenamed "Trinity." It took place in an abandoned patch of land about 40 miles (60km) away from Socorro, New Mexico, on July 16, 1945, and it exploded with the force of around 22,000 tons of TNT. It was a huge success and represented a decisive development in the Allied battle against Japan. The two bombs subsequently dropped on Hiroshima and Nagasaki, Little Boy and Fat Man respectively, were a devastating demonstration of nuclear power, with hundreds of thousands of Japanese civilian casualties. Just six days later Japan surrendered, and the Second World War ended.

HOW DO NUCLEAR BOMBS WORK?

Nuclear bombs vary in their exact working, but they all function according to the process of nuclear fission. This is where a large atomic nucleus is unstable and splits into two smaller nuclei. As it does this, it releases a large amount of energy and, crucially, a number of high-energy neutrons. The neutrons impact with other large nuclei, causing the nuclei to split, thus producing yet more energy and neutrons. This chain reaction can cause a relatively small amount of material to very quickly produce huge amounts of energy in the form of an incredibly large explosion. These explosions are many thousands times more powerful than TNT bombs, and the bombs in existence today are thousands of times more powerful than those dropped on Japan.

AFTER THE WAR

The dropping of the bomb changed the very idea of war. After 1945 nuclear weapons were produced that could likely end the world as we know it. This completely changed the nature of international politics and the ensuing Cold War had a significant impact on the entire world. Albert Einstein, whose equation $E=mc^2$ was crucial to the development of nuclear weapons, said of them: "I made one great mistake in my life. When I signed that letter to President Roosevelt recommending that atom bombs be made."

But it is not just destruction that comes from atomic research. Nuclear fission and fusion could be what we need to keep the lights on in the coming decades. Despite a few scares we are still only just starting to realize the potential.

BARDEEN AND BRATTAIN DEVELOP THE TRANSISTOR

We live in the age of transistors—tiny switches that are either off or on, "1" or "0." We interact with objects filled with transistors every day, and in terms of inventions that revolutionized the world, the transistor may just rank as the most significant. Transistors have only been around for about 70 years, but they have already changed the way we do everything.

The Second World War led to a boom in computer technology. Initially, computers used triodes—large glass valves that acted as amplifiers to electrical signals. Through the application of voltage, they could be used as a switch, turning on or off to create the 1s and 0s. They were large, unwieldy, and easy to break. Much research had gone into them, because they were useful for codebreaking and radio communications.

When the war ended, William Shockley (1910–1989) of Bell Labs began trying to reduce the size of triodes through the use of semiconductors. It was at this time that he began working with John Bardeen (1908–1991) and Walter Brattain (1902–1987). They used the electron mobility of semiconductors (explained below) to try to create a "switch" with no moving parts. They began work without much success, and the early transistors they created were highly unreliable, functioning as intended one moment and then seemingly not the next. They were also incredibly fragile and, despite various attempts with a multitude of methods, were mostly nonfunctional.

However, in December of 1947 they made a breakthrough. They took a plastic wedge and attached two thin sheets of gold to it as conductors and led wires from each sheet. From here it was found that the signal from the emitter lead would be amplified and sent out of the collector lead when a voltage was applied through the base lead (see the diagram on page 148).

HOW DO TRANSISTORS WORK?

The initial transistor design was replaced in 1951 with the far more efficient bipolar transistor. The two work according to the same principles, though the bipolar transistor is a little easier to explain.

A semiconductor is made out of three main parts, and two of them are actually the same material—germanium or, more recently, silicon. The structure of something like silicon is such that all of the atoms are fully

COMPUTER PIONEERS: Bardeen, Shockley, and Brattain (left to right) in their laboratory shortly after developing the transistor.

bound, so there are no free electrons. An absence of free electrons means there can be no conduction. However, semiconductors undergo "doping," whereby roughly one in every 1,000 silicon atoms is swapped out for the atom of another element. There are two ways to do this: adding something like phosphorus will introduce an extra electron, thereby creating a surplus of free electrons; or adding in an element such as boron will reduce the number of electrons available, leaving unbound atoms. It is easier to think of the unbound atoms as having "holes." Although these holes don't actually exist as real things, it is easier to imagine them than the whole motion of the many electrons that causes them.

Semiconductors with additional electrons are known as N-type semiconductors, because of the negative charge on the electrons. Semiconductors missing the electrons are called P-type semiconductors, because of the positive nature of their "holes." It's important to note, however, that both of the materials have no actual charge on them. However, when the two are placed next to each other, the spare electrons in the N-type will start to move over into the P-type to fill in all of the holes. At this point something called a "depletion layer" forms. This is where the electrons move over and fill in the holes but give a part of the P-type semiconductor a negative charge. The depletion layer acts as a wall and repels further electrons from passing into the P-type.

Applying a positive voltage to the P-type semiconductor is enough to overcome the depletion layer's barrier, allowing electrons to start flowing through it. What this means is that by the simple application of a voltage,

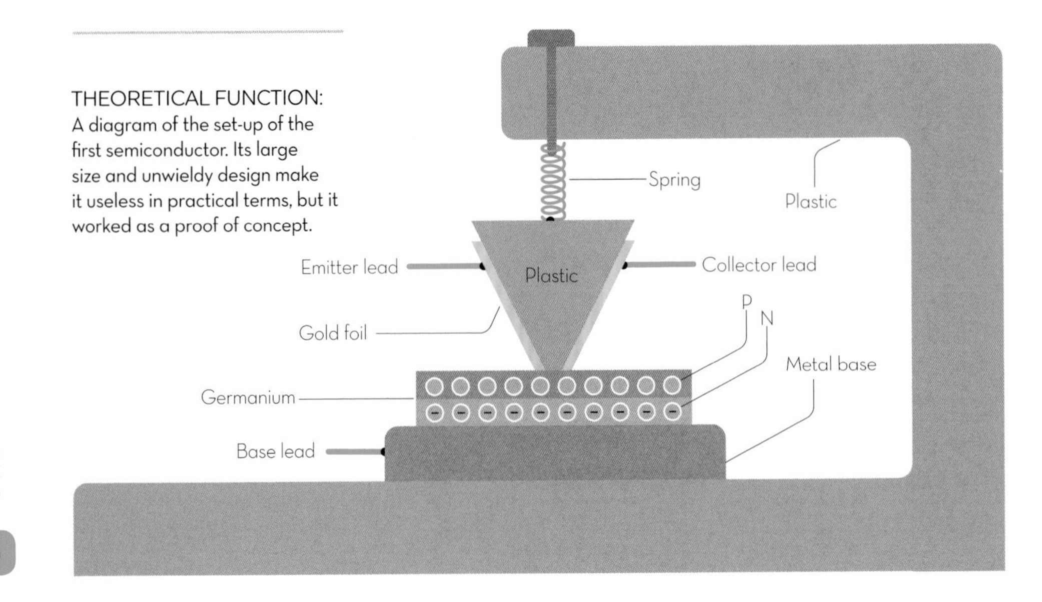

THEORETICAL FUNCTION: A diagram of the set-up of the first semiconductor. Its large size and unwieldy design make it useless in practical terms, but it worked as a proof of concept.

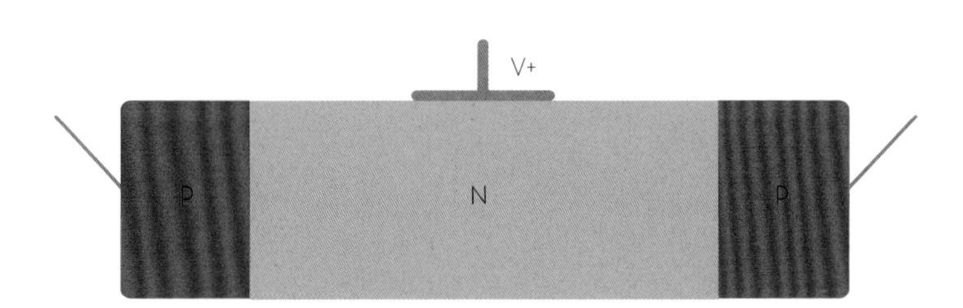

UBIQUITOUS TECH: A simple diagram of a modern transistor and how it operates. Millions of these can be found inside all electronic devices.

we can create a "switch" without a single moving part, which can be turned off or on almost instantly. This is how we use them in modern devices as 1s and 0s, whereby a turned off transistor is 0 and turned on is 1.

TRANSISTORS TODAY

The development of the transistor has made it possible to reduce the size of computers, which until that point had been extremely large. The development of ever smaller transistors led American computing entrepreneur Gordon Moore (b. 1929) to famously create his eponymous law, which states that the number of transistors that can be fitted onto the same-sized chip doubles every two years.

Today we can make transistors that are incredibly small—in the order of only a few tens of nanometers. If they get much smaller, we need to start worrying about quantum effects such as quantum tunneling, which could cause a current even in an off transistor. Their small size allows us to fit an incredible number of transistors in all of the technologies we use. From phones to computers, calculators to games consoles, everything that has a microchip in it is filled with near countless numbers of transistors. They enable the 1s and 0s of the software to be turned into a phone call to your family and a calculator to check that I got the math in this book right.

At the time of writing, Intel announced the release of a new chip that contains around 7.2 billion transistors on a surface area slightly less than that of the average postage stamp. The transistors themselves are a mere 14 nanometers in length.

THE FIRST TRANSISTOR?

In 1925, Julius Edgar Lilienfeld (1882–1963) was granted a patent for a transistor very similar to the one Bardeen and Brattain created. However, he didn't publish any papers on the device, and many of the materials mentioned were not actually available, so a working model was never built. It is for this reason that his contribution is largely ignored or forgotten, and Barden and Brattain get all the credit.

RICHARD FEYNMAN BEGINS A SERIES OF LECTURES

One of my physics lecturers once told the room, "there is no physics course that exists today that is complete without a Richard Feynman quote," and in all honesty, he was right. Feynman's lecture course of the early 1960s became the gold standard for physics teaching and continues to be used across the world to this day.

Richard Feynman (1918–1988) was born in Queens, New York, to an average white-collar American family. At the age of nine his family moved to Far Rockaway, where he attended high school and was picked up as a very bright pupil. By his mid-teens he had already taught himself a great deal of mathematics, such as differential and integral calculus, advanced algebra, and analytic geometry. He had begun work on his own form of notation before beginning a degree at the Massachusetts Institute of Technology, where he published two well-respected papers. He graduated in 1939 and, after getting a perfect score in physics on the entrance exam, began to study at Princeton University. It was during this time that he started to gain his distinctive style of breaking complex and difficult mathematics down into easy-to-understand forms. He wrote his thesis on some of the challenges posed by quantum mechanics, and some of his seminars were attended by greats such as Albert Einstein.

Even before he received his PhD in 1942, he had been recruited into the Manhattan Project, the U.S. atomic bomb project, where he worked on the computation of the potential power of atomic bombs. Although he was a relatively junior member of the project team, he was regularly approached by Niels Bohr, who would discuss problems with him. This is partly because many of the other scientists respected him too much to speak frankly, and it was only Feynman who could be counted on to point out flaws in Bohr's ideas.

On June 16, 1945, Feynman's terminally ill wife died, and he threw himself into his work, but found himself beginning to struggle. He had previously turned down a number of offers to work at Cornell University, but this time he decided to accept an offer and moved back to New York. His father died suddenly in October 1945, and he suffered a bout of depression. Following this he found himself completely unable to focus on his research, so looked to physics problems that were not so much important as they were satisfying to solve. During this time he started to develop

TALENT FOR COMMUNICATION: Richard Feynman photographed in 1959. His lecture series has become the gold standard for physics teachers.

an interest in quantum electrodynamics, even attending the 1947 Shelter Island Conference, where it had top billing among some of the brightest minds in America.

Feynman continued to work on refining his ideas on physics as well as developing his own mathematical systems. Additionally, he developed the Feynman diagram as a way of explaining particle interactions. He presented it at a conference in 1948, where it was a resounding flop. His radically new approach and the strange new diagrams possibly confused his audience, and the ideas were opposed by a number of prominent physicists, including Paul Dirac (1902–1984) and Niels Bohr. But the excellence and

THE FEYNMAN DIAGRAM

The Feynman diagram (below) is a representation of subatomic particles interacting. Its visual nature makes it significantly easier to understand than the equations that underpin it. It was this work that earned Feynman a Nobel Prize.

The Feynman diagram shows an electron (e^-) and a positron (e^+) colliding into and annihilating each other, from which the **γ** radiation is created. This **γ** radiation later decays into a quark ($\bar{q}$) and antiquark (q) pair, with the antiquark then releasing a gluon (g). The *y*-axis of the graph represents space and the *x*-axis represents time. It is worth noting that antiparticles are represented as traveling backward in time. This is due to their mathematics; they (probably) don't actually travel backward in time.

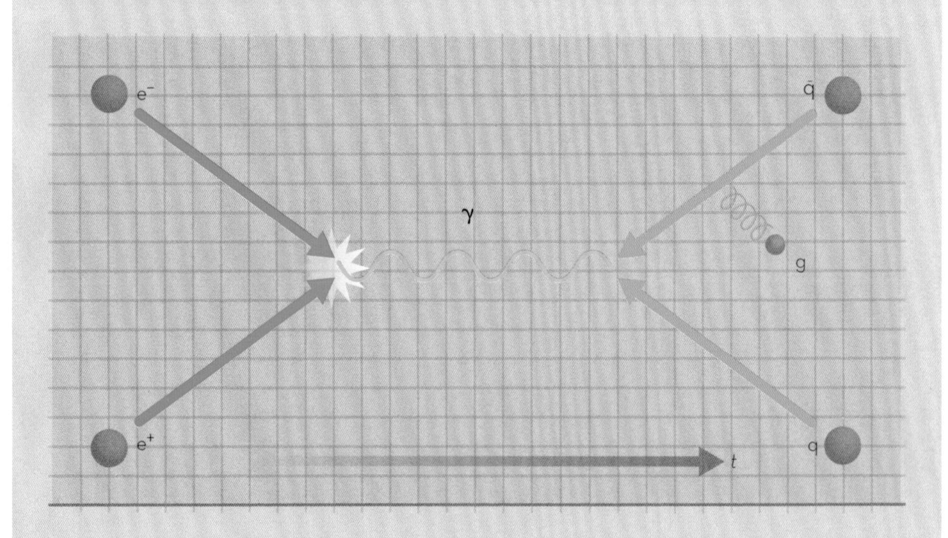

simplicity of his idea did not go unnoticed by all of the attendees, and by the early 1950s the Feynman diagram was a commonly used tool, the dissemination of which was no doubt helped by its easy translation into a format readable by computers, which allowed for it to be calculated easily.

LANDMARK LECTURES

In 1952 Feynman moved to the California Institute of Technology (Caltech), where he investigated supercooled fluids and the weak nuclear force, while also continuing his work on the Feynman diagram. He also taught physics and became a relatively popular teacher, although not yet the sparkling communicator of complex physics that he would become.

The public persona for which he would eventually be well known began to take shape when he was asked to rework his lectures. This he did, and he delivered the new series between 1961 and 1963. Feynman's style and charisma led to many of them being audio recorded, and a series of books were written based on his lecture notes. It is these books that have become one of the greatest recent works of physics. The lectures are so widely used perhaps in part because of the feeling that Feynman put into them. His entire life had revolved around physics, and he had seen it used for both good and ill. His own life had been so deeply touched by physics that the lecture series he gave was almost a personal journey.

"The Feynman Lectures on Physics," first published in 1964, comprise three volumes: *Mainly mechanics, radiation, and heat*; *Mainly electromagnetism and matter*; and *Quantum mechanics*. In the lectures, Feynman presented an excellent picture of the basics of much of physics. He was able to explain in a clear and concise manner the basis for all current research and provided the core of Caltech's degree course. Initially, the lectures seemed to fail to engage many students because they covered none of the grand, exciting new ideas of the time, but even as student numbers dropped off, the number of graduates and peers that attended grew. They appeared to find the new perspective and Feynman's clear and concise reasoning to be highly refreshing.

The Feynman lectures remain among the bestselling physics books of all time. While rarely adopted as a textbook, the Feynman lectures often serve as a basis for most university physics courses, and you'd struggle to find a single one where the students wouldn't benefit from reading it. Although the books are full of equations and not for casual reading, if you have a strong wish to understand how physics works, the Feynman lectures are highly recommended.

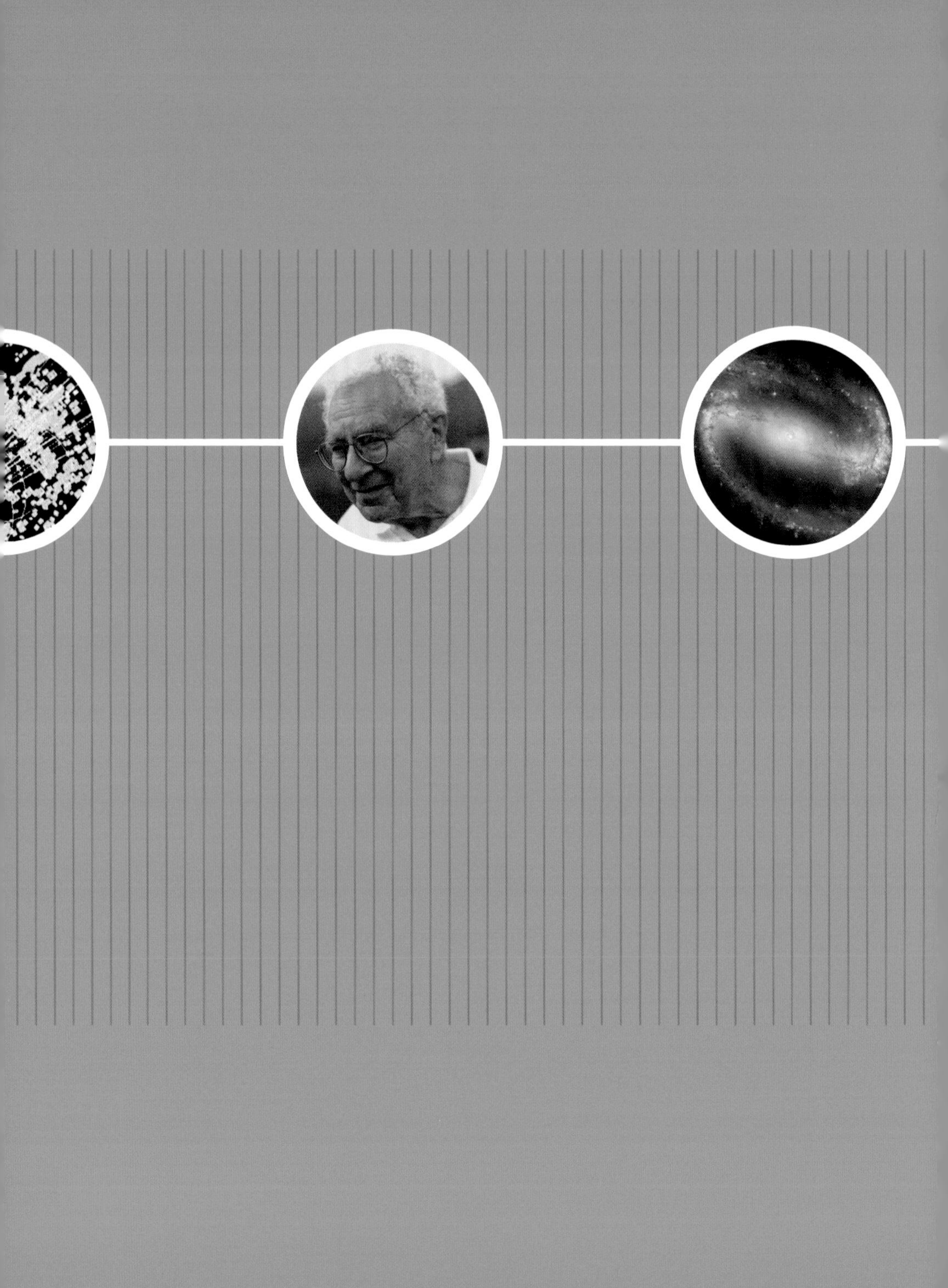

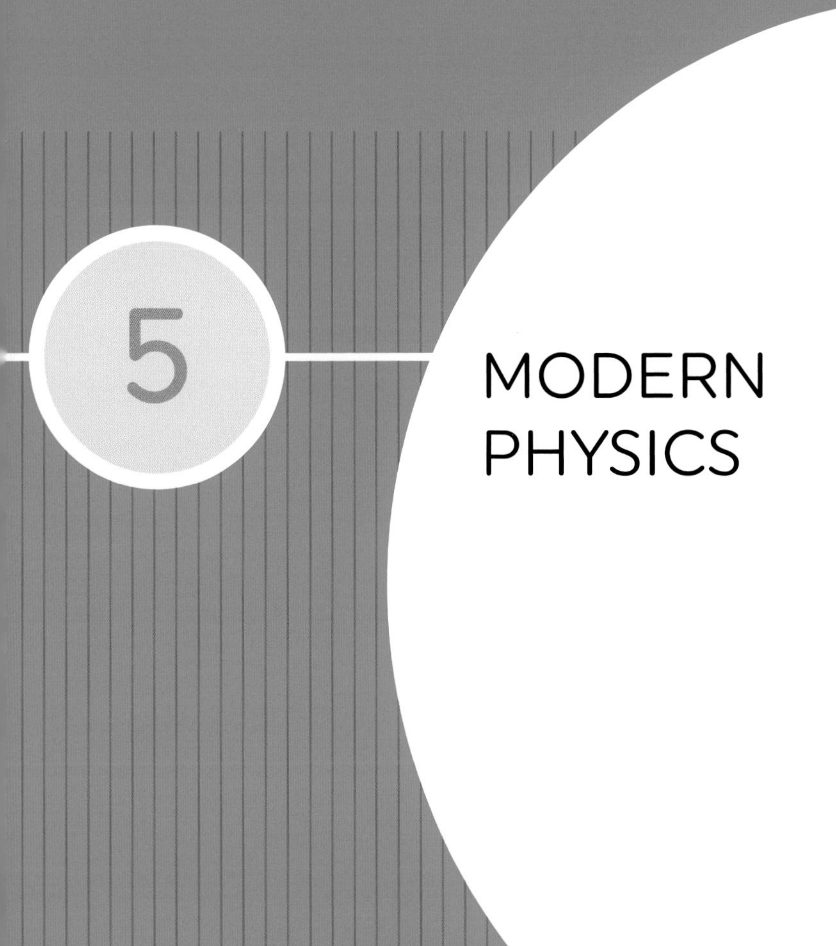

5 MODERN PHYSICS

CDC 6600 GOES ON SALE

The potential of computers had first been shown by Alan Turing (1912-1954) during the Second World War, and the invention of the transistor had dramatically increased the computing power. But CDC 6600 was something new, a machine that would revolutionize how physics is done. It was a supercomputer.

The average home computer isn't particularly efficient. Thanks to its operating system (usually Windows or MacOS of some type), the many different programs it might have, the antivirus constantly checking everything, and a million other things that the computer is constantly doing, even when it's not actually "running" anything, mean that it doesn't use all of its available power on any given task. The net result of this is that it actually runs a lot slower than it could. When we're doing high-performance calculations for physics models and tests, we can't afford to waste all that computing power, so that's what a supercomputer is for. It acts as a huge, complicated calculator.

The power of a supercomputer is measured in floating-point operations per second (FLOPS), which is a measure of how many single operations (for example, adding 3 and 2) it can do per second. An average desktop PC (at the time of writing) can produce about 7GFLOPS (7×10^{9} FLOPS), whereas the current fastest supercomputer in the world, Sunway TaihuLight, runs at 93PFLOPS (93×10^{15} FLOPS).

THE FIRST SUPERCOMPUTER: The CDC 6600 at the National Center for Atmospheric Research (NCAR) in Boulder, Colorado, in 1965.

The first supercomputer was created by Seymour Roger Cray (1925–1996). Cray joined the American firm Control Data Corporation (CDC) in 1958, where he began working on new types of computers. At the time larger computers were still built using glass triodes, but Cray took advantage of the increasingly popular transistors to make one of the first computers with no moving parts, the CDC 1604. This became the first commercially available computer to use only transistors (see pages 146–149). It sold well and CDC asked Cray to build more computers, with the focus on business use in order to appeal to a wider market. Cray, however, had different plans and directed the resources he had been given toward creating the fastest computer in the world, with the aim of building a machine 50 times the speed of the CDC 1604. He did, however, fulfill his brief by developing the requested computers alongside this main project.

Cray tried for some time without much success, finding himself limited primarily by the transistors available to him. When new silicon-based transistors and integrated circuit chips hit the market, Cray switched quickly to those and achieved a huge boost in performance, due to the fact that the new transistors could turn on and off far faster, allowing for more FLOPS. As Cray conducted his research, the company grew and became more corporate, placing greater commercial pressure on their engineers. In 1962 Cray delivered an ultimatum: the company should allow him to focus solely on his pioneering work, or he would move elsewhere. The management were loth to lose him and so accepted his demands. After this, Cray set up a lab to fully utilize the ever more reliable and easy-to-use transistors. They hit another major roadblock when they realized that the new transistors were heating the devices to dangerous levels, and there was a danger that important components would melt or even that a fire could break out. To address this, a chilled water tank was introduced to conduct heat away from the circuits. By 1964 the machine was ready to be shown to the world.

CUTTING-EDGE TECH

The CDC 6600 comprised four "arms"—large cabinets filled with the computer's components—each of which had its own cooling system and connected to a computer terminal with two screens and a keyboard. Most computers used a single central processing unit (CPU) that would load up the problem, solve it, and then output the results. Typically, a computer's performance was determined (and therefore limited) by the quality of the CPU. The genius of the CDC 6600 was that it outsourced much of the computational work to peripheral processors, which were almost separate mini-computers dedicated solely to solving the problems. This meant that whichever task the computer was engaged in could be broken down into separate parts and worked on simultaneously by the peripheral processors. This process, known as parallel computing, has become the defining characteristic of supercomputers.

The CDC 6600 worked at the unparalleled speed of 3×10^6 FLOPS, which was around ten times the speed of its closest competitor. Its market price was about $8 million and it sold over 100 units, mostly to research labs and universities. Its success heralded the beginning of an exponential increase in computing power, and the ability of computers to perform calculations at incredible speeds.

THE POTENTIAL OF SUPERCOMPUTERS

As mentioned earlier, supercomputers are essentially exceptionally good calculators. They can be used to perform many calculations in a short space of time, which makes them useful for modeling. As an example of this, imagine a ball being struck on a pool table. Taking the force with which it is struck, it is possible to calculate the acceleration the ball will undergo (using Newton's second law, $F=ma$). The equation for motion ($v^2=u^2+2as$) can then be used to calculate the ball's velocity after one second, two seconds, and so on. If the deceleration caused by friction with the table and air resistance can be calculated, it is possible to know the exact path the ball will follow when it is hit. This can be extended to the effects of its collision with another ball. Calculating this

HIGHLY EFFICIENT: A 2007 machine from the IBM Blue Gene/P series, which combines high performance with relatively low power consumption.

myself wouldn't take too much effort, but if I wanted to calculate something similar using numerous balls, it would get complicated. This is where supercomputers come in. They crunch all of the numbers for us, and provide a model of what will happen.

Supercomputers have been used for just about every type of physics research, from weather forecasting to modeling galaxies. Modeling has become fundamental to research; it is almost impossible to find a form of research that does not first use computer models to test ideas before expensive experiments are conducted. The complexity of many models requires supercomputers.

The future of supercomputers is exciting because they provide such predictive potential. Even with our current technology, we are limited; a Japanese professor used a supercomputer for several weeks to produce the most accurate model of the sun ever. When asked how accurate, he replied, "maybe... 10 per cent." So it's clear there is much work still to do. The next generation of supercomputers is expected to reach speeds of 10^{18} FLOPS, and by 2030 that will probably have reached 10^{21} FLOPS. That kind of power will enable extremely accurate two-week weather forecasts all over the world. We are only beginning to understand what will be possible with supercomputers.

THE STANDARD MODEL IS CREATED

The Standard Model encapsulates almost everything we know about the subatomic world and the fundamental forces of nature. It is the closest thing we currently have to a complete understanding of our universe.

The grand unified theory (GUT) is the holy grail of physics, an equation that could describe and link together everything and anything. To achieve this, the linking of the four fundamental forces—electromagnetic force, strong nuclear force, weak nuclear force, and gravity—would be required.

The first significant step toward this came in 1961 when the American physicist Sheldon Lee Glashow (b. 1932) managed to combine the electromagnetic and weak force together into the electroweak force. In 1964 Murray Gell-Mann (b. 1929) and George Zweig (b. 1937) independently worked on and put forward the idea of quarks, subatomic particles that combine to make protons, neutrons, mesons, and other things. Quarks were experimentally discovered in 1968, when the scattering of electrons off a proton showed that the protons themselves were made of multiple smaller hard cores. Further advances into this subatomic particle physics led to the discovery of the Z Boson (caused by the electroweak force) at the European Organization for Nuclear Research (CERN) in 1973, allowing physicists to begin building up a picture of the interconnectedness of the subatomic world.

In a 1974 paper, the Greek physicist John Iliopoulos (b. 1940) set out an early version of the standard model, in which he described the relationship between the up, down, and strange quark and predicted the existence of the charm quark (the charm quark had been speculated upon previously but he was one of the first to properly theorize it). The charm quark was experimentally discovered in November of the same year. The model Iliopoulos created also incorporated other known subatomic particles, such as electrons, muons, and their neutrinos (tiny mass-less particles made in some electron interactions) along with the bosons—the particles responsible for the fundamental forces. The photon mediates the electromagnetic force, the gluon the strong nuclear force, and the Z and W bosons the weak nuclear force.

If you were paying close attention then you may have realized something. We have bosons for only three of the four fundamental forces, which suggests that something is missing.

MURRAY GELL-MANN: He introduced not only the idea of subatomic particles but also the idea that they have some symmetry.

This realization underlines the power of models such as the standard model; using them, we are able to tell if things are missing, and also to predict some of their properties, thus potentially making them easier to find.

In 1976 Martin Perl (b. 1927) discovered the tau particle, which is a little like a heavier form of the electron. Thanks to the standard model, it was predicted not only that the tau would have its own tau neutrino but also that there would be another pair of quarks. And as expected, a year later Leon Lederman (b. 1922) theorized the bottom quark, though this wasn't found experimentally until 1995.

THE MODERN STANDARD MODEL

Today, the standard model (see the chart on the next page) looks different from that proposed by Iliopoulos, thanks to the addition of the more recently discovered particles, but it still takes on a similar form. It can be split into a number of different groups, each of which highlights different properties of the particles.

FERMIONS Fermions are the twelve particles that all have a spin value (think of it as the amount of energy it has while rotating) of 1/2. This category can be further divided into leptons and quarks. Quarks can experience the strong nuclear force and combine to make hadrons, such as protons. Leptons are difficult to detect (which is why they took so long to find), mostly due to the fact that they don't interact with the strong nuclear force, they have very little mass, and the neutrinos have no electric charge, meaning many of our traditional methods of observing small objects simply don't work. Note that the masses of all the particles increase as the chart moves left to right. Also, the particles will decay from right to left, with tau particles decaying into muons, and muons into electrons, which are stable. The same is true for bottom quarks, which decay to strange quarks, and so on. Note, however, that the neutrinos do not decay at all.

GAUGE BOSONS Gauge bosons carry the strong nuclear, weak nuclear, and electromagnetic forces. It is their movement from one particle to another that causes the force. For example, you can think of the electromagnetic force between two similarly charged particles occurring as photons being sent from one particle to another to knock it away (like a pool ball fired at another ball to move it away).

HIGGS BOSON The Higgs boson is a recent discovery (2013) that stands somewhat apart from the rest of the standard model. It interacts with anything that has mass and is what gives every other particle, including gauge bosons, their physical manifestation of that mass. Interestingly, because the Higgs boson also has mass, it must interact with itself.

NEAT SOLUTIONS

It is noticeable how well the standard model seems to fit together, with its pairs and its symmetries, and everything fitting into neat little boxes, according to a logical order that makes sense to us. Is there a reason why it should be the case? In fact, we often talk about the laws of physics, but why should there even be laws at all? Why should the constants we have looked at, such as G and c, be constant, and why are the exact values?

QUARKS

≈2.3 MeV/c^2
2/3
1/2
u
up

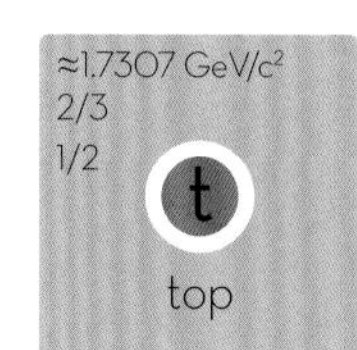

GAUGE BOSONS

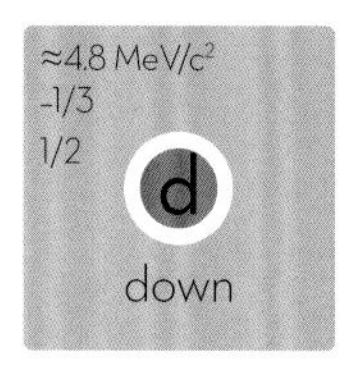

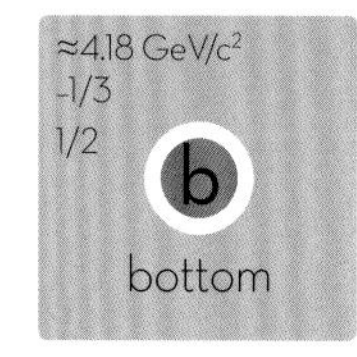

LEPTONS

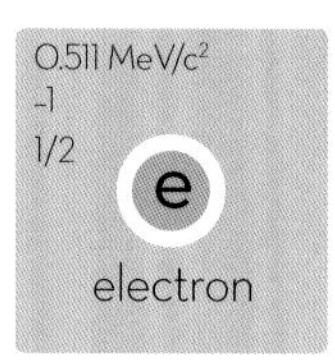

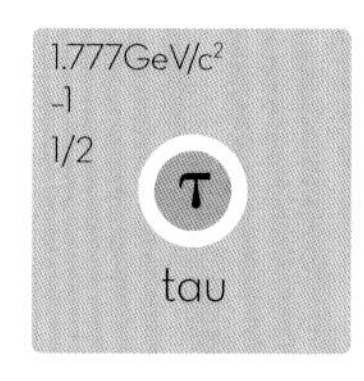

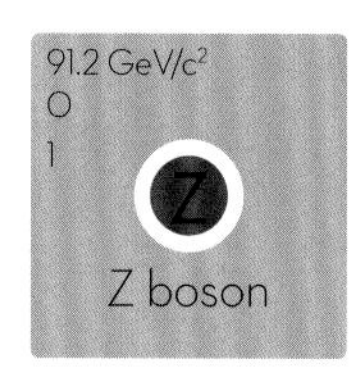

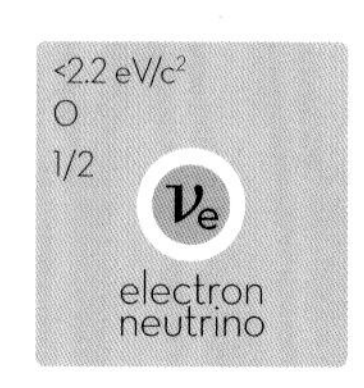

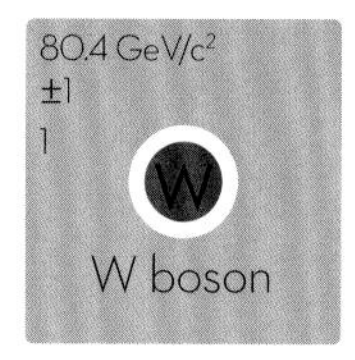

STANDARD MODEL: The standard model table is organized from left to right according to mass. The gauge bosons interact with both the leptons and quarks. The Higgs boson interacts with all the particles.

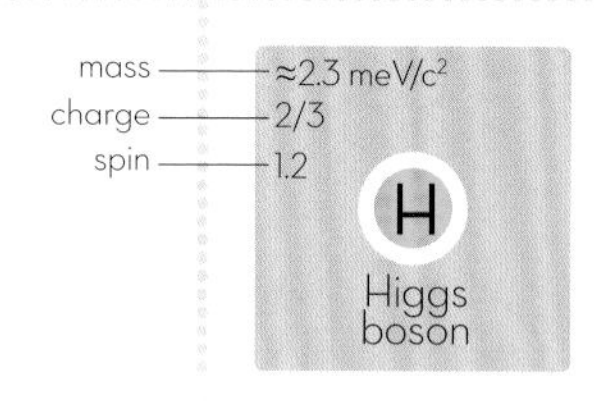

The standard model represents a new level of understanding, because this is about as fundamental as it gets. We have broken our universe down as far as we are currently able, and we still find these seemingly ordered and matching patterns there. Why does the standard model look so neat? It is not perhaps the job of physicists to answer this question, but answering it may reveal a lot about our universe, how it works, and maybe even why it does.

HIGH-TEMPERATURE SUPERCONDUCTORS ARE DISCOVERED

Superconductors are incredible materials that could completely change the way we create electrical circuits, although they present one major problem: they need to be cooled to incredibly low temperatures. So the discovery of materials that don't need such a large amount of cooling was a huge step forward.

Superconductors have a number of incredible properties, the most notable of which is that they have no electrical resistance at all. Every material has some resistance, which is caused by the electrons in the current colliding with particles of the material, thus causing some of the electrical energy to be lost as heat energy. In the USA alone, this causes an average wastage of about 10% of all electricity as it is moved from the power plant to our homes. This is a huge amount of wasted power (and money), and the figure doesn't even account for the loss caused by the wires in your own home, nor by the inefficiencies of the electronic devices you use.

The obvious answer is to make electrical components out of superconductors, and save literally billions a year in electricity costs. However, there are two major problems with this: the first is that superconductors are very expensive—a small cookie-sized disk can set you back over $100. With the potential energy savings, however, it could be a worthwhile investment. The second problem concerns what is known as the critical temperature of a superconductor, which is the temperature at which the material starts to exhibit its superconductive properties. Above this temperature, it behaves normally.

The first superconductor was discovered in 1911 by Heike Kamerlingh Onnes (1853–1926), when he cooled mercury to 4°K (–269°C). Many other elements, such as lead and tin, and a number of compounds, were also found to exhibit these properties. However, they all needed to be cooled to at least 13°K (–260°C), and to achieve this, liquid helium was needed, which is difficult and expensive to make, transport, and store. So superconductors remained a wholly academic endeavor.

RAISING THE TEMPERATURE

In 1986 two researchers for IBM, Georg Bednorz (b. 1950) and Karl Alexander Müller (b. 1927), were researching ceramic compounds. This was an underinvestigated area, because at normal temperatures ceramics are bad conductors, unlike metals,

to which the most research had until then been focused. Amazingly, Bednorz and Müller found that a compound of barium, lanthanum, copper, and oxygen becomes a superconductor at 35°K (–238°C). That is still extremely cold, but relative to the other known superconductors it was a significantly higher temperature. The discovery was a huge step forward, and it took a year for the next important discovery.

In 1987, Maw-Kuen Wu (b. 1949) and Chu Ching-Wu (b. 1941) discovered that yttrium barium copper oxide has a critical temperature of 90°K (–183°C). The significance of the breakthrough lay in the fact that liquid nitrogen was used to cool the superconductor. Liquid nitrogen has a temperature of 77°K (–196°C) and, thanks to its abundance in Earth's atmosphere, is easy to make. It can be stored in a simple vacuum flask (similar to the ones you may keep hot drinks in, only somewhat more specialized) and it is incredibly cheap to produce, costing less per liter than milk. This has made the use of superconductors commercially viable, and they have since been incorporated into technologies such as the Japanese maglev trains.

Researchers continue to search for higher-temperature superconductors. The highest critical temperature found so far, in the compound H_2S, is 203°K (–70°C), though this does need to be at 1.50×10^{11}Pa of pressure to work (normal atmospheric pressure is 1×10^5Pa). The hope is that one day a superconductor will be found that works at room temperature.

HI-TECH: A test of a maglev train in Japan. Maglevs are one of the biggest current uses of superconductor technology.

SCIENTISTS AT CERN MAKE ANTIMATTER PARTICLES

Antimatter feels like it was invented in science fiction. For every lepton in the standard model (see pages 160-164), we find an antiparticle that is exactly the same, only with the opposite charge. Should an antimatter particle and its corresponding matter particle ever meet, they will annihilate each other into pure energy. In 1995 came a breakthrough that allowed scientists to examine antimatter in much greater detail.

Antimatter exists naturally and is created (then subsequently annihilated) all the time through various processes, including types of atomic decay. It was first formulated in 1928 by Paul Dirac (1902–1984), when he showed that the antielectron (now known as the positron) was a necessary conclusion of the Schrödinger equation for an electron. The formulation of the standard model only increased confidence in the idea of antimatter. In 1932 Carl David Anderson (1905–1991) used a device called a cloud chamber to observe the tracks left behind by cosmic particles. He noted that one of the observed particles had the mass of an electron but an opposing charge.

HOW DO YOU MAKE ANTIMATTER?

Many natural processes produce antiparticles, but in order to study the particles properly

WHAT USES ARE THERE FOR ANTIMATTER?

The study of antimatter is fundamental to physics research. We can learn a lot about our universe by looking at how antimatter works and how it differs from normal matter. It may help us answer some of the biggest questions we have, such as why there is "something" rather than "nothing" in the universe, and why the universe is made out of matter instead of antimatter.

Antiparticles do already have their practical purposes. Positron emission tomography scans (PET scans) observe the behavior of the energy given off in a positron annihilation inside a patient's brain in order to diagnose a number of potential issues. While this kind of research is at an early stage, it has a number of potential applications, including in new and better materials, in new diagnostic and treatment devices, and even potentially in an energy source.

PROTON ACCELERATOR: A 1954 photograph of the Bevatron proton accelerator. It was operational until 1993.

and try to do something with them, it is necessary first to make them. The first example of antiparticle production was a device called "the Bevatron," which was built in 1954. It was a large circular device that accelerated protons so that they would collide into one another and create a shower of antiprotons. The Bevatron also produced the first man-made antineutron in 1956.

With all the constituent parts for matter already discovered, physicists at CERN set out to make some antimatter. A naturally occurring antideuteron (a single antiproton and antineutron combined) was discovered in 1965, at which point efforts were made to produce it in the lab. In 1995, scientists managed it by bombarding a xenon atom with antiprotons. Across a three-week experiment, they made nine antihydrogen particles.

One of the big problems associated with antimatter is storing it; the particles produced in 1995 only lasted for around 40 billionths of a second before annihilating. If it comes into contact with matter, it will be instantly destroyed. There are a few different methods used, the most common of which relies on using the antimatter's charge. The particle is created inside a series of magnetic fields, arranged such that the particle is pulled in all directions equally, and therefore is suspended in empty space. This method has been used to hold antimatter particles in existence for up to 16 minutes, giving physicists plenty of time to study them.

THE HUBBLE SPACE TELESCOPE TAKES THE DEEP FIELD PHOTOGRAPH

The Deep Field, taken by the Hubble space telescope between December 18 and December 28, 1995, and its successors, the Ultra Deep Field and Extreme Deep Field, are some of the most breathtaking and extraordinary photographs ever taken.

Take a pin and hold it out at arm's length. In 2003 the Hubble space telescope set its sights on a tiny patch of sky about the same size as the head of that pin. It opened up the shutter and, for about four months, collected up the tiniest amounts of light that reached it. This patch of sky was "blank"; nothing had ever been seen or detected from it. Yet when the photograph emerged, it showed an incredible array of stars and galaxies.

In the photograph, there are fewer than 20 stars (they are distinguishable by their crosslike forms, which are products of the way the telescope is built). Everything else is a galaxy, a collection of several hundred billion stars, along with nebulae, planets, black holes, and so much more. In one 24-millionth of the sky, nearly 3,000 galaxies were discovered. When this total is extrapolated, we arrive at an estimate of 7.2×10^{10} galaxies in our universe—a staggeringly large number. A further image, taken a little later and in another part of the sky, produced very similar results, giving credence to this estimation.

The Hubble Deep Field (HDF) is important because it shows us just how big our universe is. It also poses lots of questions about how and why so many galaxies form, and whether they have anything in common. Furthermore, it provided proof for two major ideas: first, it showed that the farther away the galaxy is, the more red-shifted it is, supporting Hubble's idea of universal expansion. It also showed that there are not so many local stars within our own galaxy, which became leading evidence against the MACHO theory of dark matter (see pages 138–139).

Between September 24, 2003, and January 16, 2004, an even more detailed photograph was taken of the same region of sky, revealing more than 10,000 different objects.

SEEING INTO THE PAST

Many of the objects uncovered in the HDF are so far away that their light has taken enormous periods of time to reach us. What we see in the HDF is actually light from billions of years ago. We find that many of the galaxies in it are malformed and misshapen because they have not yet settled into the spiral shapes that so many of our local galaxies have. By looking across

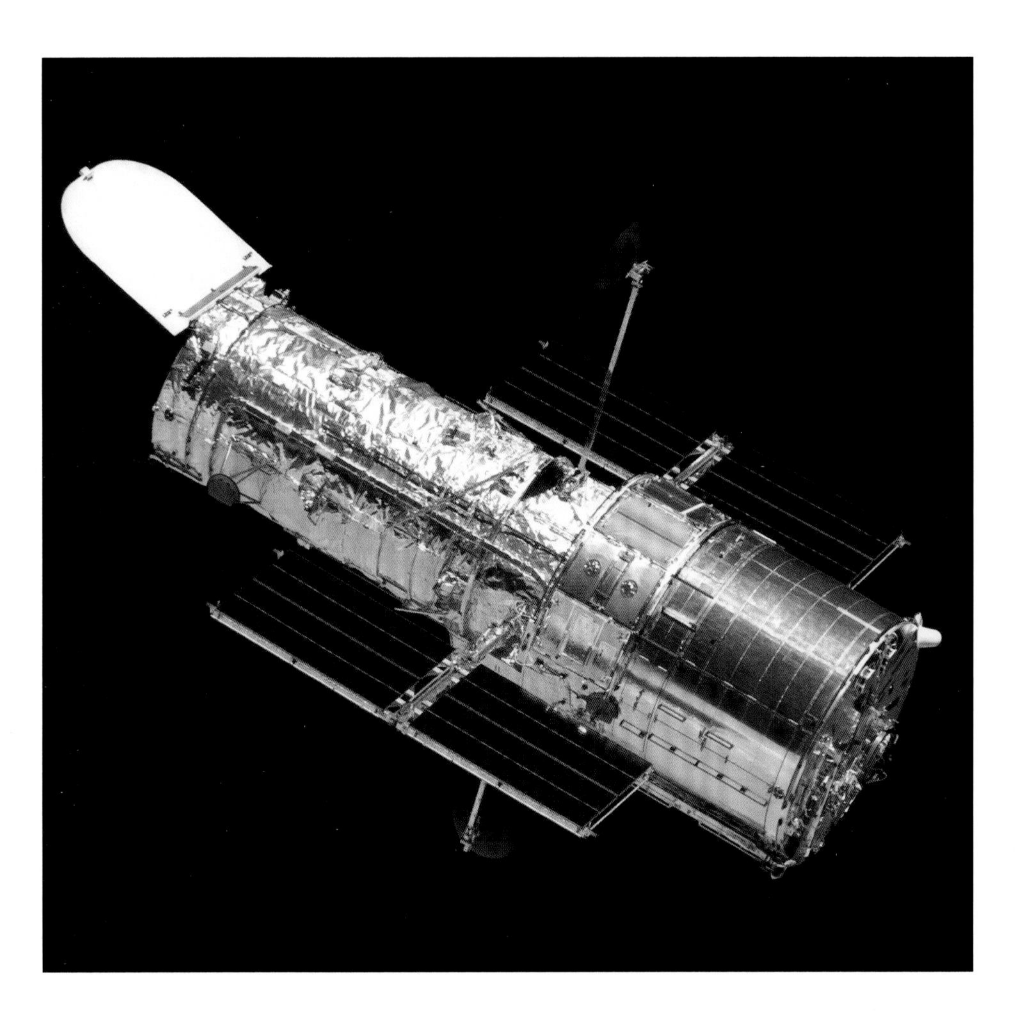

the series of galaxies in different stages of their life, physicists have been able to observe how galaxies form and move, and how many stars they produce, depending on how old they are. They have determined that within the first couple of years after the Big Bang there was suddenly a boom in the creation of galaxies. These galaxies birthed most of their stars during this time. It is now thought that many of them in fact collided into each other to form larger galaxies, such as our own and those near us.

THE HUBBLE SPACE TELESCOPE: The telescope was built by NASA with financial and technical support from the European Space Agency.

JET SETS THE WORLD RECORD FOR CREATING FUSION POWER

Fusion energy could be the future, but it's not going to be easy. It is proving to be one of the biggest engineering tasks of the modern age, although in 1997 the Joint European Torus (JET) set a world record that proved it could be done.

Fusion seems like the perfect way to make energy; it produces no greenhouse gases or nuclear waste, and its main products, helium and neutrons, can be pretty useful for all sorts of things. And we know that it's a really good source of energy—it is how the sun and all other stars make their energy and it gives 10,000,000 times as much energy per reaction than just about any other chemical reaction. So why aren't we using it already?

The main problem is that to start fusion a pretty extreme environment is required. You need a very high pressure—many times that of Earth's atmospheric pressure—to push them close together and then temperatures of over 100,000,000°C in order to give the hydrogen atoms enough energy to overcome the electromagnetic repulsion between the two. Getting these initial conditions to trigger a fusion reaction is very difficult. What's more, maintaining a fusion reaction is also difficult. While the energy required to keep it running could be gained from the reaction itself, due to anomalies in the way the hydrogen plasma works (which is not yet fully understood), fusion reactors can only be run for a finite period before they must be turned off.

JET

JET is a tokamak-type fusion reactor, which means its internal cavity (where the fusion takes place) is a giant ring in which the hydrogen is heated up and pressurized into a plasma, which is kept contained through the use of magnetic fields. JET was opened on April 9, 1984, and became the hub of much fusion research. In 1991 it achieved the first controlled release of fusion energy.

WHERE DOES FUSION ENERGY COME FROM?

As we saw with Einstein's equation $E=mc^2$, atoms have binding energies (the reason why the complete atom has less mass that its parts) and it is this binding energy that comes into play with fusion. When we take two hydrogen atoms and fuse them together into a helium atom, the change in binding energy means that excess energy is released.

In 1997 it produced more energy than it took to make it work, with a net total of 16MW (16×10^6W). An oil power station, "Ince B," which was built in the UK in the same year as JET, produced 500MW, albeit over a longer period of time. While fusion is not yet a workable solution as a clean, abundant energy source, JET at least set the world record for the most power produced by fusion, at the same time showing that fusion power is a possibility.

There is clearly still a lot of work to be done, and physicists are doing their best. The next-generation fusion reactor ITER (International Thermonuclear Experimental Reactor) is being built in southern France. When switched on in 2020, it is hoped that the reactor will produce 500MW of energy from an input of just 50MW—a net output of 450MW. This matches most traditional fossil fuel power plants in production. The hope is that by 2050 the first commercial fusion power stations will open, and we can begin to power the world with fusion.

FUSION ENERGY: A 2001 photograph of the inside of the JET tokamak. It is just under 20 feet (6 meters) in diameter and holds around 26,500 gallons (100m^3) of plasma while in operation.

GALAXY ZOO LAUNCHES

Computers are incredible research tools. They can process amounts of data far beyond what a team of scientists could handle. They do however currently have one big flaw: they are terrible at looking at pictures. So how can we process large amounts of photographic data without computers? Step forward citizen scientists.

The development of so-called "citizen science" began with the Sloan Digital Sky Survey, the most detailed complete sky map ever, which was produced via numerous mechanical telescopes, all photographing and collecting data on the sky in as much detail as possible. This created a huge amount of data, and scientists were left wondering how to deal with it all. One team was interested in classifying galaxies with the hope of learning more about how and why galaxies form in the way that they do. However, they had over a million images of galaxies to look through and classify, which is simply too many for a small research team to go through, and computers are not able to do it for them.

The result: on July 11, 2007, Galaxy Zoo was launched. This is a citizen science project that invites ordinary people to donate a bit of spare time to help classify galaxies. Anyone with access to the Internet can simply log onto the Zoo at www.galaxyzoo.org and, after a brief training session, begin classifying galaxies. Initially, contributors were asked to distinguish between elliptical, spiral, or merging galaxies (the project is more sophisticated now), and the answers would be tabulated, ready for the scientists to work with. With over a million images, and with each image needing verification from tens of people, scientists expected to have to wait years for the work to be complete. However, within a day of the launch, they were getting nearly 70,000 classifications per hour, and within a year they had hit 50 million classifications.

Galaxy Zoo continues, in updated form, to this day, and you can still go on it and help classify galaxies. The project has been used in over 48 different scientific papers, helping us to learn more about galactic formation and behavior. Volunteers working on the project also discovered two entirely new, never seen before astronomical objects: Hanny's Voorwerp (a light echo of a huge explosion named for its discoverer, the Dutch teacher Hanny Van Arkel) and Green Pea Galaxies (a type of very dense galaxy comprising a huge number of stars). Galaxy Zoo almost instantly became the most popular and well-known citizen science project and prompted an increase in the number of similar projects.

THE RISE OF CITIZEN SCIENCE

Citizen science enables almost everyone to take part in cutting-edge research, and it provides an invaluable tool for scientists. Zooniverse (the

company that grew out of Galaxy Zoo) now has 46 different projects, ranging from hunting for supernovas and tracking penguins to characterizing the surface of Mars. How these projects are set up too is changing. Now, involvement isn't limited to selecting what category an image fits into. Cancer Research UK's *Play to Cure: Genes in Space*, for example, takes information about how players chose routes through space to identify potential cancer-causing gene mutations.

The popularity of both projects and the willingness of the public to help with citizen

STARGAZING: The spiral galaxy NGC 1300, photographed in 2004 by the Hubble telescope, is one of the many millions of galaxies now visible to us.

science have led to contributions in gene studies, medicine, astronomy, zoology, and even archaeology. While computers are being trained to get better at pattern recognition, for the foreseeable future the willingness of ordinary people to help science progress is still vital. I'd encourage you to look up some for yourself and help out!

HE-RBX 02
HE-RBX 01

THE LARGE HADRON COLLIDER IS SWITCHED ON

The Large Hadron Collider (LHC) is the biggest and most expensive physics experiment ever built, designed to probe some of the most pressing questions of our day. The commencement of its experiments launched a new era of particle physics that we are only just starting to see.

Below the ground on the Swiss–French border near Geneva, Switzerland, lies 17 miles (27 kilometers) of tunnel. The bulk of the LHC is located in these tunnels. Two beam pipes 6.3cm (2½in) in diameter are surrounded by a couple of tubes 1m (3ft) in diameter containing magnets, coolant pipes, insulation, alignment detectors, and much more. At full power it can accelerate a beam of particles up to 6.5TeV (6.5×10^{12}) and then smash the particles into each other at one of the four detectors, which are so sensitive that the position of the moon has to be taken into account. These detectors feed into the largest computing grid in the world, which collects about 30 petabytes of data (1×10^{15}) per year and distributes it to nearly 200 computing facilities, which process the data and in turn send it to tens of thousands of researchers and institutes. In all, the device cost around €3.6 billion ($3.27 billion) to build and was designed and built with the collaboration of over 10,000 scientists and engineers.

THE LHC: The interior of the CMS detector, part of the LHC, photographed in 2014. The scale of the project is clear even in this one photograph.

The LHC has been attempting to answer some of the fundamental questions we have about the universe. How did the early universe work? What do particles do in such extreme conditions? Does the Higgs boson exist? It is hoped that the data from the LHC will enable scientists to explore various realms, among them the relationship between general relativity and quantum mechanics—to help push us toward a Grand Unified Theory, supersymmetry, extra dimensions, and maybe even dark matter. The LHC is a hugely ambitious effort to test the standard model, to look for all of the particles and forces we think should exist and to search for evidence of those we haven't yet theorized.

Construction took from 1998 to 2008, and in 2008 an abortive attempt was made to turn it on (the problem was a "magnet quench"). A large section of the superconducting coils had its internal magnetic field changed too quickly, causing it to lose its superconductive properties. This meant that it heated up very quickly, and six tonnes of supercooled helium flowed out of the machinery, damaging many of the magnets and the surrounding equipment.

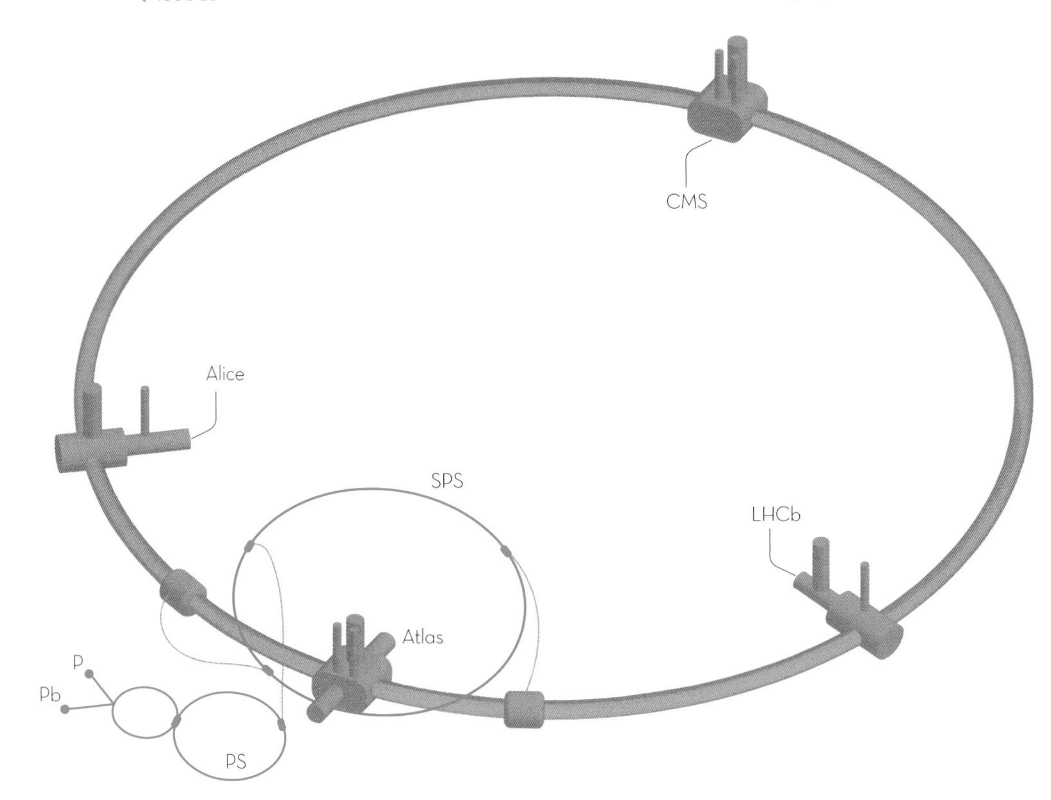

Repairs were made and a new antiquenching system was built that would cause all the excess heat to be immediately dumped out of the system into giant metal blocks, which can reach several hundred degrees during the process, thus saving the expensive materials.

On March 30, 2010, the LHC was successfully switched on, and 2808 bunches of 1.1×10^{11} protons were fired down the primary accelerators. They were sped up and released into the main beam pipes, with half moving clockwise and half anticlockwise. At this point the protons sped through the LHC 11,245 times per second. At the detectors, where the two opposing beams met, there were about 600,000,000 collisions per second, all of which were captured and measured.

THE LARGE HADRON COLLIDER: A plan view of the CERN particle accelerator, with its four primary detectors—ALICE, ATLAS, LHCb, and CMS—plus two further accelerators—SPS and PS.

WHAT THE LHC TELLS US

The biggest and most important discovery has been the uncovering of evidence for the Higgs boson. It acts as the mediating boson for gravity, much like gluons do for the strong nuclear force, or photons for the electromagnetic force. While theorized in 1964 by Peter Higgs (b. 1929), until recently there had never been any evidence for them. This changed in 2013, when scientists from CERN announced that they had found a Higgs boson at a mass of 126GeV. But this is just the beginning for the Higgs boson; there may be more types than scientists have predicted, or it might not even be the Higgs boson that has been identified, prompting us to revise our approach.

Since its launch, the LHC has also shown that an entirely new type of particle was possible. Quarks normally combine in two different ways: two quarks can combine to make a meson and three quarks for a baryon, which is like a proton or neutron. But the LHC has found evidence of strange particles, such as X(3972) or Z(4430), which seem to be made of four quarks! There is also evidence of quark–gluon plasma, which could be present just on the outside of black holes, teaching us much about how black holes work. But for all the successes, there are many theories that the LHC has not yet found evidence for. Similarly to the Michelson–Morley experiment, which was expected to prove the existence of ether, the LHC was expected to provide information about supersymmetry (the idea that bosons and fermions are paired up, and that they are the same, only with different spin values). But the verification didn't materialize, leading to the rejection of the theory by many scientists.

There is still a significant amount of data to look through, and the LHC's equipment is being constantly upgraded, with new experiments frequently conducted. There are hints at entirely new types of physics just on the horizon. While, for now, experiments conducted using the LHC seem to confirm the standard model, possible developments are looming that could be the beginning of something new. Even at the time of writing, the LHC is gearing up to test the claim of a Hungarian research team, who say they have seen a new type of boson from the decay of a beryllium-8 nucleus. If the LHC finds this boson, we will have discovered a fifth force of nature.

WORLD-ENDING BLACK HOLES?

You may remember that around the time of the switch-on a scare story surfaced about the LHC making a black hole that would destroy the world. This stemmed from wild predictions by some scientists about proton collisions potentially collapsing into black holes. Although this could theoretically happen, there was in reality never any danger. This is for two main reasons: first, the energies the LHC was using don't come close to the energies that would be required for a black hole to be formed in such a collision; second, even if they did have enough energy to form a black hole, it would be so small that it would evaporate almost immediately and would pose no danger at all.

GRAVITATIONAL WAVES ARE DETECTED

Gravitational waves—ripples in the curvature of space-time itself—were the last great prediction of Einstein's general relativity. It took over 100 years, but now they have been detected, and the discovery has opened the door to a new way of exploring our universe.

Gravitational waves are in theory created by anything that accelerates. However, gravity is a very weak force and as it travels it loses its energy very quickly. In practice then, we only expect to be able to see gravitational waves (for the time being at least) if they come from incredibly powerful events out there in space, such as two very dense objects (like black holes or neutron stars) orbiting each other or a supernova explosion.

Gravitational waves had been indirectly detected in slight variations in binary pulsar stars since 2005, but this was not evidence enough. On September 14, 2015, the LIGO experiment (see below) detected gravitational waves in a signal called GW150914. The event lasted 0.2 seconds and was caused about 1.4×10^{25}m away from Earth by two black holes, each around 30 times the mass of the sun, orbiting each other before collapsing inward and merging together. The wave had been traveling from that event toward us for over a billion years.

After verification and further study, the observation was announced in early 2016 to much celebration by the physics community. This was the final piece in the general relativity puzzle cementing its place as the dominant theory of gravity.

LIGO

The laser interferometer gravitational-wave observatory (LIGO) is based in Livingston, Louisiana. It was originally built as a joint venture between Caltech and MIT, and in 2008 brought in the UK Science and Technology Facilities Council and the German Max Planck Society to increase the sensitivity of the device through the Advanced LIGO project. In all, the project has cost around US$1.1billion.

LIGO works in a surprisingly similar way to the Michelson–Morley experiment. It consists of a 20W laser that first passes through a power recycling mirror, which focuses the light and increases it up to 700W. This is then split by a beam splitter and sent down two paths of exactly equal length (4km [2.5 miles] each). The light then reflects off a fully

PRECISION OPTICS: The optics used in the LIGO gravitational wave detectors, which comprise mirrors between which light beams are bounced.

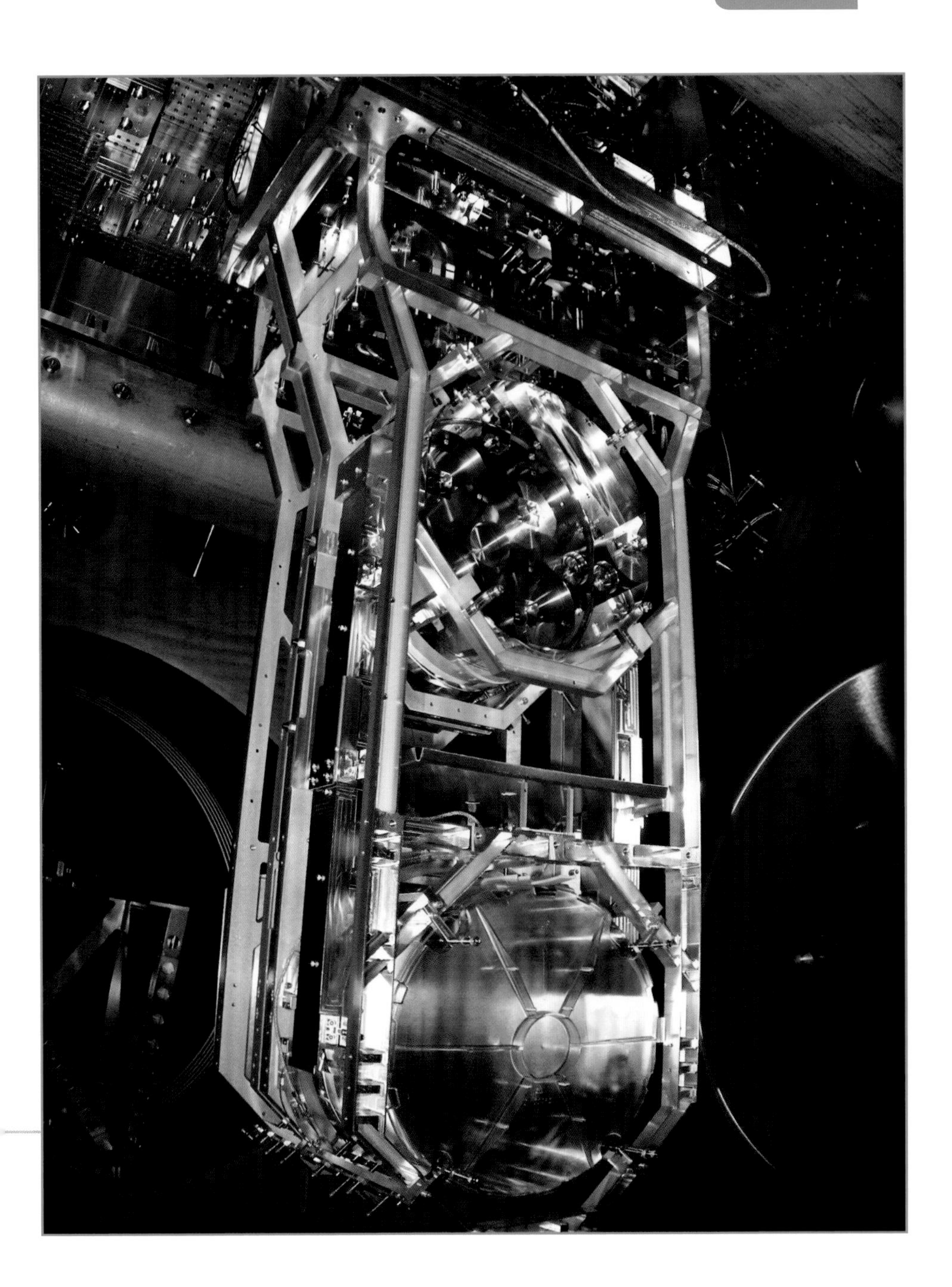

reflective mirror at the end of the arms and back down. There is a partially reflective mirror on each arm that reflects much of the light back to the other mirror, which then reflects it back to the partial mirror, and so on. The light reflects around 280 times, which makes the effective length of each arm 700 miles (1,120km).

During normal running the device is designed so that the laser passes through the arms and eventually arrives at a light detector that is exactly out of phase, which means that the two waves will interfere destructively with one another and then the detector will get no reading. However, if a gravitational wave passes through the device, then it will cause space–time to become distorted differently in each arm. This means that when the light arrives back at the detector it will not be perfectly out of phase, because one will have traveled fractionally farther than the other one, and this will cause a fringe pattern to appear.

As incoming gravitational waves are so weak, the physical change in length caused by a gravitational wave across the 4km arm is only 1×10^{-18}m—that's less than one thousandth the width of a proton. This means that the instruments need to be incredibly sensitive. However, this makes them very vulnerable to things such as constant tiny Earth tremors, and even vibrations from cars passing by miles away. In order to reduce things like this, LIGO is suspended and kept at supercooled temperatures in a vacuum. There are also, in fact, two LIGO detectors—one in Livingston and the other near Richland, Washington. As gravitational waves travel at the speed of light, a gravitational wave should cause exactly the same event in both detectors, but with an approximately

DETECTING GRAVITATIONAL WAVES AT HOME

It is potentially possible to detect gravitational waves from spinning neutron stars that aren't perfectly round. The little bumps on such a star would cause a small wave every time the star spins, and neutron stars can spin several thousand times per second. This would cause a small, constant signal. Looking for these is rather simple but very intensive in terms of computational power. So the LIGO team launched Einstein@Home in 2005. It is a different type of citizen science, requiring you to download the software to your computer, which then uses your spare computing power to run calculations. Over 300,000 people from over 220 countries have downloaded the software, making it act basically like a massive and spread out supercomputer. At times its performance has reached over 2 PetaFLOPS, which would make it one of the most powerful supercomputers in the world. You can download the software yourself and take part.

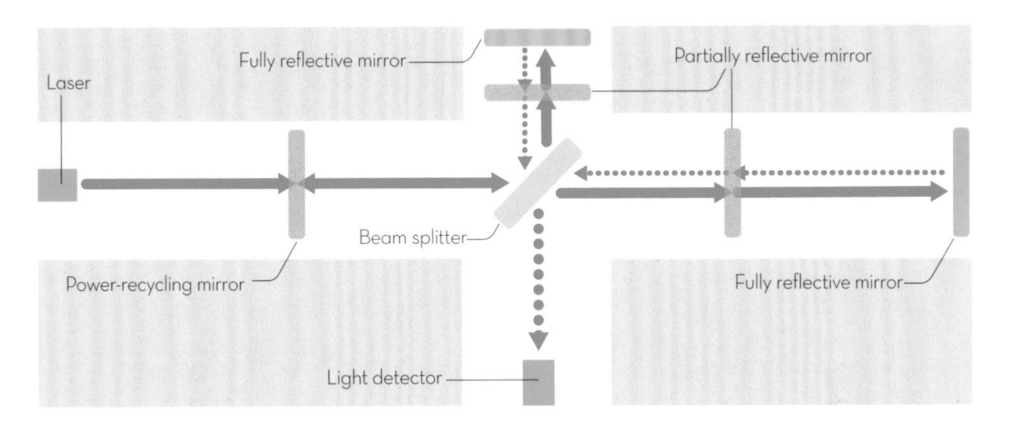

GRAVITATIONAL DISTORTION: The existence of a potential gravitational wave would mean the two light beams are no longer perfectly out of phase.

ten-millisecond (1×10^{-6}) delay, this helps confirm the gravitational waves and remove any anomalies. By using this exact time delay and some geometry, it is also possible to roughly calculate where the wave came from, allowing us to try to identify the source.

HOW CAN WE USE GRAVITATIONAL WAVES?

At the moment almost all astronomy is done using electromagnetic radiation. First scientists used their eyes, then the telescope, and later many parts of the electromagnetic spectrum, such as infrared or microwave radiation. Every time a new method of observation is used, we learn much more about the universe. Gravitational waves could be the next tool that we use to observe the heavens. Gravitational waves have the advantage of not being blocked by anything. While light might be blocked by dust cloud or planets, gravitational waves can pass through most things undisturbed, meaning that we might be able to use them to observe places we normally can't. Electromagnetic radiation also has a finite starting point called the "recombination event," which is the point at which it is first possible for photons to be produced. This means we can't use electromagnetic radiation to look any farther back than that. Gravitational waves, however, were likely around before that. By using them and studying the cosmic microwave background (CMB) created by the recombination event, we may be able to discover more about the early universe.

There is the problem that gravitational waves are very weak, meaning we would only be able to use them to look at incredibly massive objects, such as the aforementioned binary systems and supernova. However, even if their use were limited to these objects, we would still gain much insight into how our universe works. There remains a lot of work to be done and many more secrets to uncover.

FURTHER READING

Blair, David. *Ripples on a Cosmic Sea: The Search For Gravitational Waves*. New York: Perseus, 1999.

BookCaps. *Aristotle in Plain and Simple English*. Kentucky: Golgotha Press, 2012.

Bolton, Henry. *Evolution of the Thermometer*. Charleston: Bibliolife, 2009.

Bryson, Bill. *Seeing Further: The Story of Science and the Royal Society*. New York: William Morrow Paperbacks, 2011.

Butterworth, Jon. *Smashing Physics*. London: Headline Publishing, 2015.

Calinger, Ronald S. *Loenhard Euler: Mathematical Genius in the Enlightenment*. Massachusetts: Princeton University Press, 2015.

Cercignani, Carlo. *Ludwig Boltzmann: The Man Who Trusted Atoms*. Oxford: Oxford University Press, 2006.

Clarke, Chandra. *Be the Change: Saving the World with Citizen Science*. CreateSpace Independent Publishing Platform, 2014.

Clery, Daniel. *A Piece of the Sun: The Quest for Fusion Energy*. New York: Overlook Press, 2014.

Collier, Peter. *A Most Incomprehensible Thing: Notes Towards a Very Gentle Introduction to the Mathematics of Relativity*. Incomprehensible Books, 2014.

Cwiklik, Robert. *Albert Einstein and the Theory of Relativity*. New York: Barron's Educational Publishing, 1987.

Dawson, John. *Logical Dilemmas: The Life and Work of Kurt Gödel*. Florida: CRC Publishing, 2005.

Ferguson, Kitty. *The Nobleman and His Housedog: The Strange Partnership That Revolutionised Science*. Auckland: Review Publishing, 2002.

Fermi, Enrico. *Thermodynamics*. New York: Dover Publications, 1956.

Fitzpatrick, Richard. *A Modern Almagest*. Texas: University of Texas, 2013.

Goddu, André. *Copernicus and the Aristotelian Tradition*. Leiden: Brill, 2010.

Gow, Mary. *Archimedes: Mathematical Genius of the Ancient World*. New York: Enslow

Grifiths, A. B. and R. M. Wenley. *John Dalton: The Founder of the Modern Atomic Theory*. CreateSpace Independent Publishing Platform, 2016.

Heisenberg, Werner. *Physics and Philosophy: The Revolution in Modern Science*. New York: Harper Perennial Modern Classics, 2007.

Hermann, Hunger, and David Pingree. *Astral Science in Mesopotamia*. Leiden: Brill, 1999.

Hesketh, Gavin. *The Particle Zoo: The Search for the Fundamental Nature of Reality*. London: Quercus, 2016.

Ifrah, Georges. *Universal History of Numbers: From Prehistory to the Invention of the Computer.* New York: Wiley, 2000.

Kaiser, Cletus J. *The Transistor Handbook*. Florida: CJ Publishing, 1999.

Khun, Thomas S. *The Copernican Revolution: Planetary Astronomy in the Development of Western Thought*. Massachusetts: Harvard University Press, 1992.

King, Henry C. *The History of Telescopes*. New York: Dover Publications, 2011.

Krauss, Lawrence M. *Quantum Man: Richard Feynman's Life in Science*. New York: W. W. Norton & Company, 2012.

Kupperberg, Paul. *Hubble and the Big Bang*. New York: Rosen Publishing, 2005.

Kutner, Mark L. *Astronomy: A Physical Perspective*. Cambridge: Cambridge University Press, 2003.

Mahon, Basil. *The Man Who Changed Everything: The Life of James Clerk Maxwell*. New York: Wiley, 2004.

Malham, Simon. *An Introduction to Lagrangian and Hamiltonian Mechanics*. Edinburgh: Herriot-Watt University, 2015.

Metz, Jerred. *Halley's Comet, 1910: Fire in the Sky*. South Carolina: Singing Bone Press, 1985.

Murray, Charles. *The Supermen: The Story of Seymour Cray and the Technical Wizards Behind the Supercomputer*. New York: Wiley, 1997.

Newton, Isaac. *The Principia: The Authoritative Translation and Guide*. Trans. I. Bernard Cohen. California: University of California Press, 2016.

Oerter, Robert. *The Theory of Almost Everything: The Standard Model, the Unsung Triumph of Modern Physics*. New York: Plume, 2006.

Orzel, Chad. *How to Teach Quantum Physics to Your Dog*. London: Oneworld Publishing, 2010.

Pais, Abraham. *Niels Bohr's Times, In Physics, Philosophy and Polity*. Oxford: Oxford University Press, 1994.

Polkinghorne, John C. *The Quantum World*. Massachusetts: Princeton University Press, 1986.

Rhodes, Richard. *The Making of the Atomic Bomb*. New York: Simon & Schuster, 2012.

Robinson, Andrew. *The Last Man Who Knew Everything*. London: Oneworld Publications, 2007.

Sparrow, Giles. *Hubble: Window on the Universe*. London: Quercus, 2010.

Steffens, Bradley. *Ibn Al-Haytham: First Scientist*. Greensboro: Morgan Reynolds, 2007.

Stockli, Alfred. *Fritz Zwicky: An Extraordinary Astrophysicist*. Cambridge: Cambridge Scientific Publishers, 2011.

Swenson Jr., Loyd S. *Ethereal Aether: A History of the Michelson–Morley–Miller Aether-Drift Experiments, 1880–1930*. Texas: University of Texas Press, 2011.

Thomas, J. M. *Michael Faraday and the Royal Institution: The Genius of Man and Place*. Florida: CRC Press, 1991.

Whitehouse, David. *Renaissance Genius: Galileo Galilei and His Legacy to Modern Science*. New York: Sterling Press, 2009.

Wilson, George. *The Life of the Honourable Henry Cavendish*. Leaf Classics, 2013.

INDEX

PICTURE CREDITS

17 © Creative Commons | Fae
21 © Leemage
33 © Photoresearchers Inc
53 © Universal Images Group
57 © Andrew Dunn | Creative Commons
60 © Christie's Images | Bridgeman Images
65 © Getty Images
69 © Royal Astronomical Society | Science Photo Library
76 © North Wind Picture Archives | Alamy Stock Foto
82 © Science & Society Picture Library
91 © Popperfoto
92 © Bettmann
97 © Stefano Bianchetti
103 left © Science & Society Picture Library
103 right © Emilio Segre Visual Archives | American Institute Of Physics | Science Photo Library
104 © Emilio Segre Visual Archives | American Institute Of Physics | Science Photo Library
129 © Getty Images
132 © Margaret Bourke-White
135 © Alfred Eisenstaedt
137 © Bettmann
141 © Bettmann
147 © Emilio Segre Visual Archives | American Institute Of Physics | Science Photo Library
151 © Joe Munroe
156 © University Corporation For Atmospheric Research/ Science Photo Library
159 © Argonne National Laboratory | Creative Commons
161 © Joi | Creative Commons
165 © Saruno Hirobano | Creative Commons
167 © Ullstein Bild
171 © Efda-Jet/Science Photo Library
174 © Tighef | Creative Commons
179 © Caltech | MIT | LIGO Lab | Science Photo Library

All other images in this book are in the public domain.

Every effort has been made to credit the copyright holders of the images used in this book. We apologize for any unintentional omissions or errors and will insert the appropriate acknowledgment to any companies or individuals in subsequent editions of the work.